Automotive Vehicle Safety

This unique book is both a practical design guide and a valuable reference book. The information it contains is essential for specialists such as designers, engineers, manufacturers and lawyers, who need to make well-informed decisions in order to ensure automotive safety. The inclusion of discussion topics and worked examples makes the book relevant to students as well as professionals.

Automotive Vehicle Safety is an internationally-oriented discussion of how to evaluate products, processes, services and systems. The authors identify key generic safety principles and discuss their applications. Decision-making criteria are also explained, and in-depth information on human simulation, human error control and driver distractions is provided. The book details reconstruction techniques and methods of crash testing, and looks at future vehicle safety and universal design.

George A. Peters is a licensed attorney, engineer, and psychologist. He is a 25 year member of the Society of Automotive Engineers and has been elected a Fellow of the American Association for the Advancement of Science and the Royal Society of Health.

Barbara J. Peters is a licensed attorney and is a Fellow of the Roscoe Pound Foundation.

Automotive Vehicle Safety

George A. Peters, Esq., J.D., P.E., C.S.P., C.P.E., FIOSH, FRSH

Registered Professional Engineer (California, Safety and Quality)
Licensed Psychologist (Medical Board of California)
Counselor at Law (California and U.S. Supreme Courts)
Fellow, The Royal Society of Health (U.K.)
Fellow, American Association for the Advancement of Science
Member, Society of Automotive Engineers
Santa Monica, California, U.S.A.

and

Barbara J. Peters, Esq., J.D., FRPF

Attorney at Law
Peters & Peters
Santa Monica, California, U.S.A.

CRC Press
Taylor & Francis Group
Boca Raton London New York

CRC Press is an imprint of the
Taylor & Francis Group, an **informa** business

A TAYLOR & FRANCIS BOOK

CRC Press
Taylor & Francis Group
6000 Broken Sound Parkway NW, Suite 300
Boca Raton, FL 33487-2742

First issued in paperback 2019

Typeset in Sabon by Keystroke, Jacaranda Lodge, Wolverhampton

ISBN-13: 978-0-415-26333-7 (hbk)
ISBN-13: 978-0-367-39587-2 (pbk)

British Library Cataloguing in Publication Data
A catalogue record for this book is available from the British Library

Library of Congress Cataloguing in Publication Data
A catalog record has been requested

Visit the Taylor & Francis Web site at
http://www.taylorandfrancis.com

and the CRC Press Web site at
http://www.crcpress.com

Dedication

This book is dedicated to Roberta B. Peters who reviewed and constructively questioned the drafts of each chapter, typed and proofed the manuscript several times, and encouraged the two already overworked authors by stating that the information contained in this work could materially help those who are in a position to improve automobile vehicle safety.

Contents

List of figures

List of tables

Preface

This book is intended to fulfill a rapidly growing need for useful information on how best to meet the demands and achieve an ever higher level of safety for products, assemblies, systems, processes, and services. It should be rather obvious that design safety has become far more important because of the more severe consequences from failures, defects, discrepancies, recalls, improper risk assessments, and purely subjective decision-making. This is an era of increasing competition, shorter design cycles, rapid transition and change, pervasive electronics, and the delegation of design safety to many suppliers, subassemblers, and outsourced assemblies. Some of the resultant issues are: how can there be an assurance of safety with decentralized control? What are the important design and system concepts necessary for the attainment of a high level of safety throughout the design-to-disposal and recycling process? What is needed, expected, and effective? And how is it done with the existing tools of the trade?

Form

This is not an argumentative, opinionated, or advocacy publication. It does not represent the unique views of any special interest group. There are no tales of disastrous consequences, no specific product or name identification, nor any controversial speculative opinions or conclusions. Instead, the book attempts to present helpful, straightforward, reliable, and factual information that is reasonable, acceptable, practical, contains a fair balance of interests, and with an expression of cultural limitations that might be realistically expected. A balance has been attempted so that there will be minimal burdens commensurate with the desired product and system improvement, successful attainment of safety goals and requirements, and compliance with social expectations.

Application

There are many types, brands, and models of automotive vehicles to which design safety procedures could be applied. A partial sampling of the universe might include passenger vehicles, trucks, buses, motorcycles, snowmobiles, lawn and garden tractors, and other self-propelled vehicles. It might include construction equipment, such as excavators, motor graders, bulldozers, backhoes, trench diggers, and other earthmoving vehicles. It might include farmstead equipment such as tractors, mowers, and combine harvesters. It could include original or aftermarket accessories from hitching devices to electronic entertainment. It could include consideration of the effects of agricultural chemicals in vehicle mounted tanks, hazardous material in refuse trucks, and environmental factors on military vehicles. Whatever the application, there is significant communality in terms of the application of design safety techniques, concepts, principles, and methodology.

Acceptability

The contents of this volume have been discussed by the authors in lectures that have been sponsored by professional associations and government agencies. Key concepts have been published, by the authors, in peer-reviewed journals. Drafts have been reviewed by sophisticated and experienced specialists. These efforts have proved very helpful in formulating and presenting information that can be considered acceptable by peer-review publication standards. Much of the information is new in published form, but it springs from and is conditioned by actual experience. The hope is that it will serve to help others in the difficult task of improving safety for the benefit of all persons, interests, and enterprises.

Integrated safety

The authors of this book believe that the contents are an application of the basic principles of what they call **integrated safety**. This is in sharp contrast with the separate application, by various disciplines at different times and locations, of a variety of special techniques such as those that predominate in system safety, reliability engineering, human factors, quality, risk management, loss control, behavioral safety, traffic and transportation engineering, intelligent vehicle research, accident reconstruction, and, in particular, design engineering.

Gender, dollars and measurements

In the text that follows, measurements are given in English units, with some conversions to metric units where it might be helpful. English units are the most commonly used in the automobile industry, but there is a gradual shift toward the International System of Units (SI); there are many symbols and terminology that could be used, and even the notation for kinetic energy may be in any one of four symbols. For conversions and precision, we recommend the use of references such as *The CRC Handbook of Mechanical Engineering* (1998), 19–2 to 19–11 (F. Krieth, ed.), Bacon Raton, Fl: CRC Press. The monetary units are given in US dollars, which may be converted into any currency as of a given date.

For simplicity, the masculine word 'he' is meant to apply to both sexes. It also reflects an industry heavily populated by males. However, it is a fact that some very talented females now occupy positions in every aspect and corner of the industry, from entrance level occupations to chief executive officers of corporations. This growing sexual diversity is accompanied by cultural diversity resulting from the assimilation of various national groups into the increasingly international operations of various enterprises.

Acknowledgements

The authors wish to express their appreciation to those publications and organizations that permitted an author's reservation of book rights or otherwise made this volume possible. This includes, in whole or in part, publications and lectures from the *Journal of System Safety*, the *Journal of the Royal Society of Health*, the *Journal of the International Society for Technology, Law and Insurance*, the Human Factors and Ergonomics Society, the National Research Council of the National Academies of Science (USA), and the Society of Automotive Engineers.

Disclaimer

The authors and the publisher must strongly disclaim any and all liability that could conceivably result from the use of the information, in whole or in part, contained in this book. This is because of the unique circumstances that may arise during the application of such information, the wide variation in competence and capability of users or appliers, and the need for highly skilled specialists. Before any detrimental reliance, the reader is urged to obtain consultation, advice, and opinions from appropriate licensed professionals. This is particularly important when decisions and actions are being taken that are within the realm of a licensed professional practice. Decisions on safety issues often involve matters relating to several disciplines such as a branch of engineering, medicine, and the law. What each specialist might perceive, understand, and recommend may be based on limited data that is material to human health and safety. Every application is unique, novel, and requires balancing of a variety of interests and alternatives.

1 Introduction to vehicle safety

(a) Objectives

The purpose of this publication is to provide useful information that could save lives, prevent personal injury, reduce property damage, and generally improve the individual's quality of life.

It is assumed that a better understanding of good design practices will enable product improvement that manifests significantly less risk to humans, machines, and the environment. A better comprehension of the overall requirements may help to reduce system errors and faults. A broader appreciation of societal interests may help in balancing risks and determining what is reasonable under the circumstances. In some respects, this document could serve as a complementary 'safety design manual' or supplemental 'safety training handbook' for a wide variety of industrial and commercial enterprises. Its academic importance should be self-evident.

(b) Good intentions are not enough!

Despite the good intentions of many engineers, safety problems regularly occur in a wide variety of products, processes, and systems. If safety were merely a matter of good intentions and common sense, there would be few if any accidents, recalls, liability fears, lost profits, or problems of insurability, loss of use of cash reserves, and possible adverse publicity that could affect market shares. Design safety is not achieved by chance or hope, complacency or compliance, the application of ethical and moral values, or simple exhortations as to exemplary safety objectives. Achieving safety is a fairly sophisticated process and requires special effort.

We have entered a new era of fairly constant design improvements, more radical changes in design, ever more compressed design-to-market cycles, high-volume production schedules, more diverse marketplaces, and significant world trade competition. Safety problems can suddenly appear and have dramatic consequences. Assurance of acceptable levels of product safety cannot be left to chance and the good intentions of individuals or companies. Specific techniques and assurance procedures are corporate health essentials.

(c) Adequacy of knowledge

A design engineer may have had very little academic preparation or on-the-job training in terms of specific design safety principles, techniques, or knowledge. The engineer may be suddenly assigned an important design safety function in connection with his work. There may be little or nothing available in terms of sources of information. The company safety guidelines, engineering manuals, or policy documents may abound with good-sounding generalities, but fail to address specific questions or provide the necessary help. The design supervisor or reviewer may have had little more in terms of design safety experience and knowledge, except for some highly specialized but very limited applications. The engineer may try to the best of his capability, but a lack of relevant knowledge or available tools of the trade may convert an otherwise capable engineer into a mistake-prone learner. This book may furnish some guidance and awareness that could serve to compensate for the lack of knowledge that is so prevalent in many industries.

(d) Someone else's responsibility

During the design process, the focus is usually on iteratively obtaining desired product performance, correcting problems as they arise, meeting strict schedules, obtaining design approvals, attempting to meet all specifications and requirements, and seeking to obtain customer satisfaction. Under such circumstances, safety may be assumed to be adequate and relegated to a secondary service function. It is also assumed that if there are potential safety problems, they will receive attention, testing, and correction by others. In some companies, safety is considered almost as a non-design or after-design function. In essence, design safety may be delegated, relegated, or simply overlooked in a large compartmentalized company bureaucracy. It may be a broadly shared responsibility, superficially implemented by all, with no real accountability.

An example of 'shifting responsibility' may be found in an engineer's opinion as to the 'safe use' of a product. The eventual product user is generally urged or expected to apply common sense in dealing with the product, to follow directions and instructions correctly, to avoid misuse and abuse, and to exercise due care for his own safety. This shifts the responsibility to the consumer who, logically, should act with great care as to his own personal safety. If there should be effective safe-use communication with the consumer, user, operator, worker, or bystander, it could reduce accidents. But history has shown that it will not eliminate accidents and sometimes is virtually ineffective. This is the reason why a reverse shifting of responsibility has occurred. The engineer may be considered to have more pertinent knowledge and be in a better position to design-out problems (hazard prevention) to completely eliminate the prospective source of injury. In essence, the reverse-shift is urging the engineer to exercise greater care in the design of a product, process, or service.

The lesson to be learned is that design safety is a sophisticated task, there should be clear responsibility for it, and it should not be diffused as always being someone else's responsibility.

(e) The hear-no-evil problem

During the design process, those who convey 'good news' are very welcome. This is particularly true if it involves meeting a difficult schedule milepost, passing a key test, achieving a higher than expected performance level, or coming in under budget. Those who convey negative messages usually suffer the fate of the unwelcome messenger. Unfortunately, the discovery of a possible safety problem is 'bad news' because of its negative effect in terms of required problem resolution efforts, added cost, and time delay consequences. Since design managers do not welcome bad news and this may be perceived as a desire to hear no evil, the design engineer may not want to frustrate the managers by conveying bad news about safety problems. In essence, it is politically unpopular to discuss a safety problem unless there is overwhelming credible evidence as to its actual existence, and even that may be vigorously challenged. This suggests that the design safety process be established and conducted in such a manner that it cannot be neutralized, fudged, or subjectively contorted to produce only good news. The philosophical approach should be that early discovery of a potential safety problem saves the considerable time and cost that would have to be expended at a later date.

(f) Generic implications

This book may seem to focus on product safety, but the principles and techniques are actually generic and have wide potential application. The obvious emphasis is on automotive vehicles, but this could be considered illustrative in character and analogous to what might be applied in other industries.

Vehicle safety may seem to be exclusively a design engineering function. But, as this book clearly reveals, vehicle safety problems may originate from damage during fabrication, assembly-line errors and omissions, and flaws during manufacture that are intentional in character. Safety problems may be caused by damage during vehicle transport to the dealer, particularly for import vehicles. Dealers may introduce problems during showroom demonstrations and rides, or from service and repair discrepancies. The owner-operator may subject the vehicle to misuse, abuse, or a failure to understand proper care. Aftermarket accessories and customizing of vehicles may affect vehicle safety. Thus, examples of causation from a variety of sources and relevant remedies are a hallmark of this volume.

The achievements of the automotive industry surround us and have fundamentally changed the world in which we live. The benefits are obvious. The resulting safety problems are well known and some are highly publicized. There are now many more companies in the automotive industry, they are more widely dispersed geographically around the world, they produce many more diverse and complex products, and the market for a greater variety of self-propelled vehicles in more countries is increasing. This increases the probability that more safety problems will emerge, and these may challenge objectives relating to corporate and social responsibility. Therefore, this book, which provides a comprehensive review on how to prevent vehicle safety problems, is needed by engineers, managers, and corporate officers. Safety assurance is not good luck; it must be earned by attention to detail.

2 Basic concepts of vehicle safety

(a) Underlying principles

In approaching any safety issue, there are some underlying principles that could provide valuable *guidance* in terms of initial problem-recognition. They may furnish some helpful *perceptions* related to a possible risk situation. They also assist in defining probable *needs* and in the formulation of an effective problem-solving *plan*. What is to be done, how it is to be done, and what is to be accomplished depends on how the situation is defined within a relevant knowledge and value system (a reference base). Some specialists seem to know immediately exactly what should or could be done, because their thought processes are positively guided by underlying principles of safety. In essence, understanding the full meaning of the following key principles is a necessary precursor to all other safety activities.

(1) *The public health analogy*

Vehicle safety can be conceived as a public health concern when the overall injury and fatality rate appears to be comparatively high. The emergency care physician makes diagnoses and provides for medical treatment and care for automobile accident injuries, so there is direct contact between the practice of medicine and the source and flow of automobile accident patients. One of the objectives of preventive medicine is to reduce injuries from all sources, particularly when injuries and diseases seem to be associated with a particular industry. This may be one reason why some emergency room physicians have been chosen to become administrators of the National Highway Traffic Safety Administration in the United States and automobile accident problems are widely discussed in medical journals throughout the world.

A public health approach carries with it the concepts, language, techniques, and perspectives both of medicine and of allied health professionals. The focus is not just occupant protection, crashworthiness, and pedestrian safety. It includes topics with which public health specialists are familiar, such as the effects of toxic chemicals, particulates such as asbestos and silica, and other materials. They are alert to harmful-to-health material found in vehicles, vehicle assembly plants, converter and private label facilities, and automobile dealer showrooms and their service, parts replacement, repair, and rebuilding activities.

In essence, a public health approach is of much broader scope than the focus of most automobile design engineers, whether they are primarily concerned with components, vehicles, or traffic systems. Some of the language of public health has permeated design safety engineering, but some of the logic, reasoning, and objectives remain a difficult analogy for many to comprehend. However, they constitute a vital portion of this book.

(2) Prioritization of effort

In dealing with vehicle safety problems, past experience has clearly shown what is the most effective and economical in terms of problem-solving remedies. A priority has emerged in terms of objectives, although each problem tends to be somewhat unique and to require more than one objective. The *first* priority is to find a design alternative, improvement, or remedy. The *next lower* priority may be given to safeguards or add-on safety devices, features, enhancements, or protective equipment. For example, instead of changing the design, protective barriers or covers may be attached. A *lesser priority* would be warnings designed to foster informed avoidance behavior. These are generally reserved for hazards that constitute what is called residual risk and applied only after design and safeguarding has achieved all that is reasonably possible. The *least effective*, but sometimes necessary, priorities are operating instruction, personal training, and technical communication guidance. They are the least effective because they attempt to influence or control human behavior that can be inconsistent, with rapid loss of memory, and subject to mistakes and errors. They are costly because they are not permanent, but need renewal, updating, prompting, and incentives for relearning. However, there are situations where detailed instructions and technical guidance may be necessary; for example, troubleshooting for infrequent malfunctions and component failure.

To be *avoided* are calls for safe use, avoidance of misuse and abuse, and to be careful. For example, encouraging automobile drivers to operate cell phones in a responsible manner has little if any effect on the driver distraction problem. Exhortations to act in a safe manner, rather than a careless and reckless manner, are so general and ambiguous as to be meaningless. The questions might be: exercise care about what hazard? What avoidance behavior is required? And what protective devices are recommended?

In essence, *design* provides a more permanent product or system improvement, *safeguards* can help but may be removed, overcome, or they may add other problems, *warnings* are reserved for a last-resort attempt to reduce risks, *instructions* are transitory and often overlooked, and *safe use* may serve liability avoidance purposes, but is generally ineffective because it is so vague, equivocal, obscure as to the real causation of a safety problem, and capable of being ignored. This prioritization rule and its exceptions will be found in applications throughout this book.

(3) The significant trilogy

One productive method of conceptualizing an approach to safety analysis is to think in terms of a hazard–risk–danger trilogy. The procedure starts with the identification of hazards. A hazard is that which could possibly cause harm. The hazard is then evaluated in terms of risk. Risk is the amount or severity of harm that could be caused by each hazard. The risk is then categorized as to whether or not it exceeds some established criterion as to the acceptability of the risk. This decision-making relates to the danger posed by the risk that is presented by the hazard. If a danger exists then there is an unsafe condition in the process or defect in the product. One illustrative definition of a danger is excessive preventable risk.

The objective of this approach is first to locate, then to take action to prevent or eliminate the hazard. If that is not possible, find a means to reduce the risk to a level that would be acceptable. Any appreciable remaining risk should be communicated to those who might be exposed so that they can take action to evade or mitigate the risk.

It is generally believed, in this approach, that design changes are of paramount importance, that add-on safeguards are less effective, and only as a last resort should there be special training requirements, instructions for use, or warnings.

This approach also gives priority to preventive action (preventing the problem before sale or use) rather than corrective action (attempting to correct problems after sale or during use). For further information see Chapters 3 and 5.

(4) Cause and effect

One of the most debilitating and misleading concepts is that there is only one cause of most accidents, a widely held belief. Parsimony in explanations is desirable. Moral judgment may be commendable. But quick blame provides little help in the search for effective preventive remedies, whether it is a proactive or corrective search for causation or cause and effect. An open mind is essential during the fact-finding stage of an accident investigation, since hidden causes often appear during accident reconstruction, and a substantial factor in preventing repeat accidents may be discovered only by which remedies are acceptable, capable of being implemented, and serve actually to reduce injuries and deaths. This book tends to emphasize the benefits of the *rule of multiple causation*. This means that there may be several causes that could produce an undesirable effect. All causes may need treatment or only one. One or all causes may be cured. Also, any one, or all, could result in a collateral improvement in the desired level of non-safety performance.

(5) Immediate objectives

There are certain key concepts that, if well understood and properly exploited, could play a substantial role in properly developing safety plans and achieving safety goals. During implementation of those plans and seeking appropriate applications for the techniques and 'tools' of the specialty, they provide theoretical or overall guidance for enhancing safety where there might be ambiguous specific requirements. These are important concepts that, if appropriately comprehended at an early stage, provide a platform for fully understanding the purposes and objectives of all the subsequent applications. They can furnish the derivative immediate objectives that can facilitate appropriate comprehension of what needs to be done and to help to avoid misunderstanding and the use of ineffective remedies. For example, the concept of *informed consent* can illuminate the proper pathway for an effective communications program. This is essential if other people may reasonably assume, trust, or rely upon the premise that there has been appropriate disclosure of all significant health and safety risks that might endanger them.

The concepts relating to an *expectations test* for customer satisfaction and an acceptable risk level are fundamental to appropriate safety plans and activities. The concept of preventable and *correctable human error* avoids superficial and meaningless remedies. The concept of *visualization* is key to understanding the mental processes important for accident prevention. The *hazard–risk–danger rule* is an excellent example cited by many specialists, that, when actually understood, helps greatly in safety analysis problem-solving. In this book, there are many such key concepts that can provide or infer immediate safety objectives. Such concepts do require special attention, repeated reading, thoughtful consideration, and some experimentation to find what works best in a particular situation with special needs.

(6) *Justifiable reliance on conclusions*

Each technical, engineering, or scientific discipline has its own version of the best scientific process, what is appropriate experimentation or testing, what is good peer review or sufficient general acceptance, and what constitutes adequate verification or validation. Each discipline or occupation has its own version of a utilitarian goal and a social value system. The essence of it all may be achieving the truth through repetition or providing an adequate objective basis so others can avoid a possible detrimental reliance. Stated another way, a factual foundation for action is preferable to speculative assumptions that may provide nothing of actual value or, worse yet, a harmful result.

This book contains information that could be of great value in terms of achieving a higher level of safety and health. Hopefully that information has an objective or factual foundation, and there is a sincere attempt to provide deference to other viewpoints. But conclusionary statements may be made as to the relative usefulness of specific concepts, techniques, protocols, methodologies, and applications. Differences in opinions may exist and such differences may serve to initiate action that could help clarify and resolve conflicts. But a key principle found throughout the book is an emphasis on attaining an objective and factual foundation that provides a justifiable basis for reliance on conclusions that may affect the health and safety of other people.

(b) Fail-safe

An example of a fail-safe device in the construction industry can be found in one model of a self-propelled scissors lift. Loss of hydraulic pressure is a frequent maintenance problem, and hydraulic pressures in the range of 400 psi to 3000 psi are used to actuate the lift cylinders of the vehicle. Since this equipment can be used in rough terrain and on slanted (inclined) surfaces, braking is important. Hydraulic pressure cannot be used to apply the brakes, because the hydraulic pressure frequently drops or fails to maintain an adequate force to keep the brakes energized or locked. Instead, the wheel brakes are spring-applied and the hydraulic pressure is used to release the brakes. In other words, if there is a hydraulic failure, the service brakes are actuated to stop motion and put the vehicle in a park or emergency position. The brake system is designed to be fail-safe for a critical safety function.

It seems like good common sense that if a component, assembly, or system should experience a failure or malfunction, this should not occur in a situation where the failure itself could create a high risk. But the fail-safe concept is often forgotten or overlooked in engineering design. For example, an electronic ignition module had an unacceptably high failure rate, the failure mode often occurred on busy highways, and the vehicles would stall and come to a stop in the middle of high-speed traffic. It simply failed in an unsafe location. There was no advance warning or any limp-home capability in the vehicle that might make it, arguably, a safe mode of failure.

A good example of fail-safe design is the run-flat tire. If the tire deflates, either from a foreign object puncture (blow-out) or from an air-out (slow leak), a moving vehicle might go out of control, despite some ability to move to the shoulder of the road on the rim of the wheel. The run-flat tire provides a better ability to control the vehicle and an opportunity to drive some distance for assistance. A good run-flat tire might obviate the need for a spare tire, particularly if there is some form of wireless communication in the vehicle to call for help. Even an emergency assistance telephone somewhere on the side of the road might reduce the risks as perceived by the driver of a vehicle with a run-flat tire.

As a general rule, no component or device should fail in a manner that presents or creates a hazard. Electronic devices should not fail and ignite flammable substances such as fuel or plastic materials. The steering column should not fail by becoming separated from the remainder of the steering system, thus permitting the vehicle to go out of control. The throttle assembly should not get stuck in a full acceleration mode without a quick means to return it to the idle position. The questions may be: should a device be designed to fail in the off-position or the on-position, in an activated or deactivated mode, or be located in a position where a short circuit and electrical arc could trigger an explosion?

In construction, agricultural, and military vehicles a failure should not affect the function of associated equipment. Neither should trailer, bed-mounted, or connected equipment pose a risk to the host vehicle if there should be a failure or malfunction. The independent function of each should be protected by fail-safe design functions.

If there are predictable and likely failures or malfunctions, they should not create any additional and unnecessary risks. In essence, each subassembly should be designed to fail-safe. It should be examined for its effect on critical safety functions to determine if it might fail in an unsafe manner or create unnecessary risks. If a safety problem could occur, the design should be modified so it will be fail-safe. There should be no reliance on others, such as drivers, to counteract any inherent risk that could flow from the failure of an element of the vehicle system. Fail-safe is a safety concept that is basic to vehicle design.

(c) Alternative design

It is difficult, costly, and risky to make changes after detail design and development testing has been completed. If manufacturing capability costs have been accrued and production schedules established, it is even more difficult or virtually impossible to make substantive changes to the design of vehicles. Therefore, significant design considerations and evaluations should occur very early in the design process. Alternative designs are usually ignored later in the design stage in favor of minor corrective patchwork modifications.

The design process is often cut-and-patch, test-and-tweak for retest, try a gold-plated version and make follow-ons to match, and use as many as possible old designs and parts that have been verified by actual field experience. In this context, new designs are disfavored because trim changes and marketing prowess can mask less than innovative design.

It is a basic general rule that for every design there is an alternative design. If a design appears problematical, particularly in terms of safety, then there should be a search for alternative designs. Doubts, questionable comments, hard to decide issues, poor test results, and customer uncertainties all imply that alternative designs should be considered for new or near future models of the product. Some manufacturers spend a great deal of money and time on product research and design evaluation, yet are timid about accepting favorable results, they require further confirmation or schedule the changes for some future unknown date (the keep-it-on-the-shelf basis, just in case it is needed).

There are always questions as to overdesign versus underdesign, what is needed or can be ignored, or does it cost money or save money? However, design safety is now given some priority, assuming that there is an objective basis to justify an alternative design. In essence, higher levels of safety may require alternative designs to overcome potential problems, but the alternative designs require early analysis and appropriate verification. For example, questionable designs have passed from platform to platform and model to model, despite alternative design concepts that could remedy known safety problems, and this has continued until market forces necessitated adoption of alternative designs. Early implementation of alternative designs can save brand identity and market share as well as consequential costs. Safety is now a positive inducement to consider early adoption of relevant and beneficial alternative designs.

(d) Redundancy and derating

When there is a critical safety function that involves a somewhat suspect pathway to successful operation, it may be advisable to consider *redundancy*. This is to design so that there is more than one means to accomplish a given function or task. Each (all) must fail before there is a system failure. Generally, block diagrams of the system are constructed, a mathematical model is prepared for the pathways and components (blocks), time-dependent relationships considered, and the effects of the most likely failures determined. A standby duplicate component or duplicate circuit may be used where there are weaknesses in order to increase confidence that the system will work as desired when wanted. For example, a spare tire is a *standby* redundant element; a multiple engine aircraft has engines on active redundancy for continued flight if an engine should fail. Redundancy may be built into electronic circuits to prolong service life, so that they continue even with some malfunctions or failures. In essence, redundancy may be advisable to prevent system failure that could have catastrophic results.

The stressors that should be evaluated include shock, vibration, pressure, humidity, cycling, electrical, thermal, and environmental factors. Based on these stresses, adjustments should be made in terms of anticipated failure rates. Failure interactions between parts should be considered. To reduce failure rates, *derating* is often used. This means operating the part below its rated value. In effect, this lowers the stress levels and prolongs the life of the part. Generally, the more a part is derated the longer it will last.

In essence, techniques such as redundancy and derating can serve to protect against catastrophic malfunctions and failures by providing an alternative means of continued function and prolonging service life. They may provide a soft cushion or additional time to stop after the first failure occurs.

Safety problems can be created when elements of a system are overstressed by operation above their rated value. This is just the opposite of derating. It can occur during an emphasis on lean manufacturing, overall cost savings for the company, cutting weight to increase fuel economy, simplicity of the product, and reduction (combination) in the number of parts to be contracted, transported, inventoried, and assembled. If sufficient care is not exercised by the design engineer to maintain robustness and adequate derating, the vehicle or product has a greater potential for defects. For example, one model vehicle had a suspension system appropriate for typical roadways and bumpy surfaces of poorly maintained highways. But even a small pothole or a sharp curb impact would throw the front wheels seriously out of alignment. The overall suspension system had been weakened in the quest for fuel economy and a rapid acceleration capability. The repair shop mechanics openly discussed the problem, their supervisors enjoyed a brisk realignment business, the brand name suffered, and the manufacturer did not seem to known that a problem existed.

Some people claim that redundancy can go too far. For example, tail lights help prevent rear-end collisions at night and a lamp failure might increase the risk of a collision. Therefore, two lights were added, then more, then good reflectorized lenses, until a high-style distinctive cosmetic effect was achieved with many marker lights and lenses. Brake lights were increased to three on some automotive vehicles, but on some trucks it grew to a bank of brake lights, allegedly in the name of increased conspicuousness and safety but often simply to give a new look and profile to an expensive commercial vehicle. Some truck-trailers have so many lights and reflectors that they seem designed by decorators in a holiday mood, although too many lights are certainly better than too few for approaching vehicles on the highway.

For further information, see Chapter 3, section (b)(9), 'Design models', and section (d), 'Combining risks'.

(e) Fault tolerance

It is often desirable to have a system that will continue to operate despite failures of components (such as a sensor), circuits (such as a short), or software (such as transient glitches). Certainly, it is very important to keep safety-critical systems 'up and running', including drive-by-wire, steer-by-wire, brake-by-wire, and throttle-by-wire systems. Similarly, collision avoidance, adaptive cruise control, and other 'smart systems' that could independently control the vehicle may be considered safety-critical.

The question may be just how reliable, robust, self-correcting, or fault-tolerant it is appropriate and necessary for a system to be. For example, adoption of a time triggered protocol might assure that signals are received only if they occur at a synchronized event or prescribed time and that an open slot always exists for important messages. It could be important to include an operator advisory (for a time-delayed functional failure) or warnings (for immediate functional problems).

Caveat: There may be some confusion between the terms fault-tolerant, fail-safe, and redundancy. A clarifying example might be found in tire failures. A fault-tolerant tire may continue to be used without repair or replacement, it tolerates faults without loss of normal function. A run-flat tire may be fail-safe, but it still needs repair or replacement to return to full normal functioning. A redundant tire is an extra or backup tire that can preserve normal vehicle functioning, but its paired twin still needs repair or replacement.

(f) Safety factors

(1) Design for uncertainty

During design, an extra allowance should be made to accommodate *unexpected* factors, loads, or conditions, as well as the *expected* (the reasonably anticipated, predictable, and foreseeable). Expected conditions include normal wear and tear over the stated service life of the product or until disposal, the effects of corrosion or adverse environmental conditions, and what might be called misuse and abuse. In other words, something extra is usually necessary to preserve the integrity of the product over its service life. In aircraft design, the safety factor may be small where added weight might be critical. But this minimal approach assumes that the low safety factor is justified by the past history of the component or assembly, its unique attributes and variables, and a diligent quality control effort. When there is little in the form of test results *and* prior in-field user experience, a larger or more robust safety factor is desirable. In essence, it is a matter of judgment, based on the risk involved and the supporting data, which could yield an informed design decision.

For some consumer products, such as automotive vehicle accessories, there may be a 'duty rating' on a label or specification. For example, ladders may be described as Type 1, heavy duty, capable of supporting a 250 pound load. This workload is for the man, tools, and materials, but may be misconstrued by the consumer as having a load capability for safely supporting persons up to 250 pounds in weight. That duty rating is based on a safety factor of 4, as specified in a trade association standard (ANSI A14.5–1974). Various test requirements may be specified to assure compliance with this minimal trade standard, such as the amount of allowable deflection of a structural member under a defined load applied at a specific location and in a certain fashion. However, such tests may be selected on the basis that they can be quickly and easily accomplished on relatively inexpensive equipment, and yield telltale information on a suspect variable. The tests may not include all functional loads, static and dynamic, or the actual effects of service life degradation. An appropriate service factor, for each condition of use, should be applied to compensate for any lack of specific relevant information. There may be different safety factors for different components in a single system.

(2) The disappearing safety factor

Automobile tires may be overinflated and this could result in a burst bead failure. It is common for tire manufacturers to use a safety factor of 8; that is, the bead strength exceeds by a factor of 8 the strength requirements needed to resist a normal inflation pressure for a new tire. Field experience has shown that the actual safety factor may be reduced to much less than 4 as it ages and suffers conditions that weaken the tire. For example, in fitting a tire to a wheel, additional stress concentrations may occur if the tire is mounted on a distorted wheel. There may be additional stresses if there is a lower bead hang-up on the rim's safety hump or a mispositioning on the edge of the bead seat. As a general rule, aging and corrosion will gradually decrease the remaining safety factor on many products.

(3) *Identifying the component safety factor*

A generous safety factor may be assigned to one safety-critical function for one component, but it may be unnecessary and costly to do the same for other functions or components. Thus, there may be any number of safety factors for the components of an assembly. However, there are circumstances where a general safety factor may be warranted for the preservation of the output function of an overall device. What is important is the process by which critical safety functions are identified, how the magnitude of the safety factor is assessed, by what engineering design and quality control tests it is determined or assured, and the means by which it is safeguarded during its service life or time to disposal.

Suppliers or vendors of parts often attempt to lower costs, either on their own initiative or as a result of the pressure exerted by higher tiers in the supply chain (those who purchase the parts and incorporate them into their assemblies). To lower costs, they may make changes in design, material, manufacturing methods, packaging, or shipping. They may believe that the specified design safety factors are sufficient to permit some decrease in the name of economy (desirable cost savings). Intrusion into the margin of safety may occur shortly before or anytime during a production run, so some procedure should be instituted to preserve the safety factor. Otherwise, the safety of the product may be jeopardized before it even leaves the factory.

(g) Objective appraisals

Safety will always suffer if it is approached by excessive optimism, guesswork, speculation, high hopes, and opinions unsupported by fact. There is always the temptation and inclination to minimize potential problems when there is a great deal of personal pride by the design engineer in the functionality of the product, when the engineering directors must sell the design to company management and marketing, and when there is a need to show progress on a design project. Thus, there should be a quest for objective rather than subjective information concerning possible safety problems.

One objective method is to force a systematic engineering analysis of each part, component, subassembly, assembly, and the total system. Just a detailed evaluation of a part, with a safety focus, might reveal a problem needing correction. This is a hazard prevention approach early in the design process. The method should rely on measurements, accumulated data, and objective analyses. This approach has been called system safety engineering and now encompasses about 100 separate techniques (system safety handbook). The quantitative or probabilistic approach is evident in the objective probability analysis shown in Table 2.1. In essence, uncertainties are reduced by specialized analysis, in a system context, using as much quantitative data as possible. Qualitative data may be reduced to a quasi-quantitative or forced category form of data. The process emphasizes design improvements at each analytic step and stage as well as an overall life-cycle system risk evaluation. As much as possible, there is an attempt to convert soft or unmeasurable data into something more objective and to focus attention on such areas of uncertainty and unknown risk. For some systems, these may be an embedded software safety certification, verification, and tracking requirement. A team approach is usually beneficial to assure that there is a lack of individual bias, the use of interdisciplinary knowledge, and some compatibility with other related analyses. Objectivity also requires clarification of ambiguities in requirements, specifications, and other technical documentation relating to safety.

Table 2.1 Objective probability analysis

Probability	Characteristic (risk severity)	Consequences		
		Injury	Loss	Environmental input
1×10^{-1}	Negligible (low)	No lost workdays	$2,000 to $10,000	Minimal damage
1×10^{-2}	Marginal (medium)	One or more lost workdays	$10,001 to $200,000	Mitigation possible (no law violation)
1×10^{-3}	Critical (serious)	Disability	$200,001 to $1,000,000	Reversible damage (law violation)
1×10^{-6}	Catastrophic (high)	Death	More than $1,000,000	Severe irreversible damage

Adapted from MIL-STD-882D, which is commonly used on military and civilian system safety programs in many countries.

In interpreting the objective probability analysis table, it should be understood that the criteria are for guidance purposes and may be tailored for a particular product, process, or system. For example, there may be a probability estimate labeled 'frequent' (more than ten mishaps during the planned life expectancy of the system). The use of a matrix permits a risk evaluation (severity times probability) that can be used to determine what level of authority within the organization or procurement structure can grant acceptance of the risk. Low risks might be approved by a design supervisor, but high risks might require the approval of a chief executive officer. The essence is an attempt at objectivity and design oversight.

The attempt to reduce risk to something 'as low as reasonably practical' (ALARP), as used in some European regulations, may be considered as arguably opposed to an objective analysis. This is because it may be used as an excuse; for example, what is reasonable and what is practical in general terms. It may tend to breed an acceptance of risks that could be reduced since they may appear to some to be reasonable. It may create a general sloppiness in the design culture, since it might be conducive to a climate of rationalization. It also induces a failure to challenge optimistic safety predictions based on subjective opinion.

(h) Unloading the driver

The results of overloading the driver can be confusion, mistakes, and collisions. An overload can result from just too much information to be processed at a given time, multitasking at an inappropriate time, excessive skill levels required, and too many choices for effective decision-making. Since humans are essentially single-channel information processors, this limitation should be uppermost in the design engineer's approach, not the maximum load that could be carried by a good driver or operator.

Early construction equipment sometimes involved a variety of levers, pedals, knobs, and switches together with ongoing judgments as to cable strength, boom position, loads, and locations. The sound of winches and drums, together with visual signals from ground spotters might be important. Tactile sensations as to load, pull or push, and tipping were significant considerations. The operator required considerable experience to operate the equipment without personal injury or property damage.

Each year has shown a remarkable simplification of controls and displays, reduction in vision and auditory demands, and increased productivity in both construction and industrial machines. Human error has been dramatically reduced by unloading the driver or operator.

The automobile driver now requires less skill in basic vehicle operation. But there is a trend for adding a variety of electronic communications and entertainment devices which increase the driver load.

A specific example of driver overload is the speedometer in which the pointer and numerals are in a color that about 10% of the male population has difficulty reading because of anomalous color perception. At a critical moment in night driving on a high-speed highway, the driver might have to stare-and-study the speedometer to determine the vehicle's exact speed. While this represents only a few seconds of eyes-off-the-road, much can happen when traveling at 88 feet per second (60 mph) on a poorly lighted roadway. The driver overload (excessive time) is easily corrected in the original design. The design objective of unloading the driver is becoming ever more important.

See Chapter 7 for a detailed explanation of driver overload.

(i) Positive guidance

Horse trails, carriage wheel tracks, and gravel pathways needed little in the form of markings or signs. In an 1893 automobile, each headlight contained one wax candle as the source of light, and the light fixture had glass openings to the front and side. The thin rubber-covered carriage wheels required a relatively flat and firm surface on which to run. In a 1911 automobile, there were two rounded reflectors at the rear and vehicle speeds reached up to the 40 mph range. Something better was needed in the form of roadway networks. Gradually, there was a proliferation of paved roadways, a vast increase in traffic flow composed of many types of automotive vehicles, and ever higher speeds of the vehicles. Something special was needed for highway safety.

After an accumulation of roadway signs, traffic lights, speed zones, and other attempts to improve highway safety, the cardinal rule of positive guidance emerged. The vehicle driver should be provided with the *guidance* information needed for driving decisions, when needed, and in a manner that commands attention. It should be presented in a clear and *positive* manner rather than ambiguously or with non-uniformity. The objective was to assure a safe trip in an automotive vehicle for *all* users of the roadway, including pedestrians and bystanders.

Since vehicles may travel in opposing directions or enter a roadway from the side, clear travel pathways had to be defined. Pavement markings included center stripes to separate traffic flows in opposite directions. Each lane of traffic in the same direction was separated by longitudinal pavement markings. Botts Dots (circular white raised markers) and RPMs (rectangular raised pavement markers) were added to create a sound or vibration to alert the driver when a vehicle departed from a lane of travel. Some RPMs reflected green for traffic in the correct direction and red for an incorrect travel direction. The right edge of the pavement was also marked (using paint or reflectorized overlay) to keep traffic on the paved surface and off the shoulder of the road, particularly in fog or when the driver could be blinded by the headlights of oncoming traffic.

A compact passenger vehicle was traveling fast on a four-lane rural highway. The highway was unlighted with no edge markings. The driver approached a slight curve in the road just beyond the reach of his headlights and then drifted off the pavement edge, because the road was not straight ahead as the driver believed. Unfortunately, there was a sharp drop of 2 to 3 inches between the paved surface and the gravel shoulder. When the driver attempted to steer back onto the road, with the left front tire rubbing against the drop-off, he overcorrected. The vehicle veered to the left, started going across the roadway diagonally, so he attempted to straighten the vehicle with a quick right steering maneuver. The left front wheel rim dug into the road surface and tripped the vehicle into several counter-clockwise rollovers. One unbelted passenger was ejected and the driver suffered roof crush injuries. Obviously, highway safety had been compromised by a failure to maintain the shoulder at the same level as the pavement, and a failure to mark the edge of the lane in a conspicuous fashion. The vehicle may have had suspension deficiencies and roof crush problems. The driver reacted too aggressively and may have failed to keep a focus on the direction of his lane. In conclusion, there was insufficient positive guidance and multiple causation as is common in such accidents and injuries.

Shoulders should be suitable to serve as deceleration lanes, for emergency parking, and should be approximately level with the roadway (AASHTO, 1976), but this is often not the case and is condoned for cost reasons. The driver expectancy may be that it should be a safe use area.

Lane markings should provide a clear unobstructed pathway for the vehicle. For example, a small pickup truck was traveling at a moderate speed in a rural area on a two-lane unlighted roadway at night. He guided the vehicle by the centerline pavement marker. He was unaware that a narrow bridge with 2.5 foot high reinforced concrete walls (railings) encroached several feet into his lane. About 3 inches of the right front of the vehicle collided with the right wall of the bridge. There was no edge marking, no narrow bridge sign in advance of the obstruction, and no delineators to outline the narrowing of the vehicle pathway. There was no end treatment (object marker) on the wall in terms of reflectorized diagonal markings or attenuating crash cushion devices for a softer ridedown (longer vehicle crash pulse for survivability).

Guide signs provide information to drivers in a simple and direct fashion. They may be route identification, destinations, or exits. There may be warning signs of railroad crossings, speed limits, intersections, deer crossings, advance notice of traffic control devices, soft shoulder, reverse turn, no passing zone, dead end, or stop ahead. In the placement of warning signs, advance notice (location) may be determined by the posted speed limit or the 85th percentile vehicle speed, and the PIEV time (perception, identification and understanding, emotion and decision-making, and volitional and execution time). The PIEV time generally ranges from 3 seconds for a general information sign to 10 seconds for 'high driver judgment' situation (*Manual on Uniform Traffic Control Devices*, 1988, at 2C–2a).

A passenger vehicle was traveling at a high speed on a state highway in a forested area. As he rounded a curve the driver saw a major intersection and a stop sign. There was no advance warning sign. Although he quickly depressed the brake pedal, it was too late, he came to a stop in the intersection and was struck by an oncoming vehicle. The questions were whether his brakes were deficient in some way, would he have slowed in compliance with an advance warning, or did he have a very slow PIEV time?

The positive guidance concept has found widespread use beyond safe highway design. Wire-guided vehicles are used in automobile assembly plants, with the lanes clearly demarked for the safety of pedestrian workers and visitors. Tests have been run on intelligent vehicle systems, with automobiles following designated pathways, properly spaced, and without driver control. For troubleshooting, repair, maintenance, and servicing scenarios, the operator traverses a conceptual pathway, is channelized or kept in the proper lane to a desired destination, may experience traffic control delays, is kept from diversions and obstacles, and is provided with timely guidance information, directions, and instructions. The key concept is to keep uppermost the need for step-by-step guidance information in traversing a pathway through the unknown or even a somewhat known situation.

Despite positive guidance on the highway, there is still a need for collision avoidance devices, route navigators, embedded road edge communicators that signal warning devices within the vehicle, and radio warnings of temporary emergency situations ahead in the roadway system. As vehicles develop, the highways should improve to accommodate them in a safe manner.

For further information, refer to Alexander and Lunenfeld (1999), Baerwald (1976), AASHTO (1973), and Chapter 5 in this book.

(j) Warnings and instructions

Warnings are based on the premise that it is only fair to alert an individual to the presence of a hazard that could injure him or cause damage. The warning should provide a fair opportunity for those exposed to take action to protect themselves from harm. Instructions provide guidance to assure proper equipment performance without unnecessary and preventable harm.

It is important to recognize the principle that warnings are a type of informed consent; that is, once informed of the risk the individual has a choice. He can avoid the risk or by implication consent to voluntarily assuming the risk. The second principle is that warnings should only be used for residual risks; that is, for the risks that cannot be reduced or minimized because of technical, economic, or practical considerations.

Instructions may be necessary to guide human behavior and achieve the desired human performance. It is important that the instructions be available when and where needed, they should effectively communicate accurate and helpful information to the user, operator, or consumer, and they should be in a form that is so simple and straightforward that they will be accessed when needed. For further information, see Chapter 5.

(k) Shielding (guarding)

Access denied may be injuries forestalled. Injury prevention may be well served by rendering both hidden and obvious hazards less accessible to undesired human behavior. Many hazards could be shielded or guarded. The rotating fan blades used to generate air flow over the engine and through coolant devices (radiators) seemed to be an obvious hazard, but injuries occurred (primarily to fingers of persons attempting service, maintenance, and repair operations). A shroud or shield was placed around the axial perimeter of the blades, serving as an accident prevention guard while improving the air flow.

Guards have been used to shield power take-offs, universal joints for agricultural equipment, and in the form of insulation for high voltage wires in automobiles and automotive vehicle equipment. Unfortunately, guards may not be as simple and effective as they at first appear. In fact, they are often the result of gradual evolution to correct or improve continuing safety problems.

Covers, shields, and guards are often removed and forgotten. Thus, they may need to be hinged (permanently connected) and be held firmly in the open and the closed positions. They may need to be interlocked, so that repair work cannot be performed during machine operation or inadvertent actuation of the machine by others. A jog or limited movement option may be necessary for troubleshooting, maintenance, or repair operations. Access denial should be sufficient to overcome the ingenuity of the poorly informed and macho inclinations of the self-reliant car lover.

The following example is instructive if the reader believes protective guards are just a matter of common sense. A safety shield was advertised for universal joints in the 1952 *Red Tractor Book*. It included a shield for the tractor power take-off shaft, one for the implement gear box, and one for the universal joint in the middle between the power take-off and draw bar hitch. It advised that the entire shaft was to be guarded or shielded. The shields were designed in accordance with the American Society of Agricultural Engineers' (ASAE) and the Society of Automotive Engineers' (SAE) standards. The shields had applications that included cornpickers, small combines, binders, windrows, balers, harvesters, and beaters. It identified a safety problem: the removal of the shields several times a day to grease the joints, and a failure to replace the shields. Some 40 years later tractors drawing farm implements were often found without safety shields for their universal joints. They were discovered because of clothing entanglements on snags (projections) of the rotating shafts and joints, with severe injuries to farm workers. The question was often: who was responsible for installing the drive line guards and keeping them in proper condition? Was it the tractor and implement owner, the original equipment manufacturer, the distributor of the equipment, or the seller of the shafts and universal joints?

It is interesting to note that the first standard on agricultural tractor power take-offs was adopted by the SAE in 1924 and it indicated that 'the tractor should be equipped with a power take-off master shield' with a strength sufficient to support the weight of the operator. It also stated that the 'manufacturer of the driven machine shall provide adequate shielding for that part of the power line drive which he furnishes'. It would seem that the responsibilities were clearly indicated, more than 75 years ago, but the arguments continued for many years. Such safety standards, particularly for guards and shields, should be interpreted in such a fashion as to prevent accidents without delegation of that obligation to others.

Some 50 years ago, shields were often used to protect crops rather than people. For example, a high-speed cultivator of small plants had a crop guard to reduce the risk of 'smothering'. The question may be: what is most important in the eyes of the purchaser?

There has been dramatic improvement in farm equipment in recent years and it is increasingly rare to find safety deficiencies such as a lack of guarding. The large equipment manufacturers have design safety manuals requiring engineers to pay particular attention to the need for guards and shields and how best to design them. But surprises never end, when equipment of all sorts arrives on the farm from many manufacturers in many countries. The same can be said of commercial, industrial, and consumer equipment.

(l) Interlocks (event sequencing)

It may be desirable positively to prevent the improper, untimely, or inadvertent actuation of a particular function. This might be achieved by requiring some sequence of events to occur before the function or event can occur. The final event is locked-out (interlocked) until the precursor (sequence of) events take place. For example, the automotive vehicle cannot be started unless the transmission is in the park position, rather than start in a forward or rearward gear position. Otherwise, a turn of the ignition key might cause the vehicle to unexpectedly crawl forward or lurch rearward.

The parking (formerly emergency) brake is rarely used by automobile drivers. They tend to use the park position of the gear shift lever to lock the wheels when parking. Sometimes, the vehicle may be left in gear or in neutral as the ignition key turns off the engine. The unattended vehicle could move or be moved. To prevent such a hazard, an interlock has been used to require the transmission to be in park before the ignition key can be removed.

In terms of distracted driving, the map navigator display may require lengthy eye fixation and some priority in cognitive processing. To prevent such high-risk driver distraction, interlocks may be used to prevent navigator use while the vehicle is in motion or only when pointed away from the driver.

The interlocks or event sequencing are safety devices that control undesirable or potentially unsafe behavior. They may force a desired sequence of events for dealer troubleshooting equipment, specialty vehicle controls, vehicle operation in difficult environmental conditions, or for aftermarket and accessory devices in construction, mining, fire fighting, or military operations.

(m) Facilitate avoidance behavior

There are situations where the demands on the driver exceed his intrinsic or built-in capability. For example, the distance a driver might see at night may be limited by the reach of the automobile headlights. Unfortunately, the vehicle may approach an object hidden in the dark. While the driver might avoid hitting such an object by lowering his approach speed, it is well known that vehicle speeds are increasing while human visual capability remains unchanged. To extend the driver's sight distance, a heads-up display (HUD) has been used to visualize what lies ahead in the dark. An infrared beam penetrates the darkness; the return signals (sensor and computer converted) are used to generate a reflected image that is projected on the windshield. The driver can now see in the dark and avoid hidden objects and anticipate curves in the road.

More intense white illumination from the headlights might seem advantageous, but the adverse side-effect might be glare and the temporary blinding of the driver of an oncoming vehicle. High-beam and low-beam headlights have another side-effect, a police enforcement problem as to whether the headlights are in the dim or low-beam position when required and whether the headlights are properly aimed.

In essence, there are many situations where human performance should be assisted or could be enhanced. There may be a vehicle design means to aid a person in order to facilitate the avoidance of hazards. It is often argued that residential street lights and highway illumination are enough, but this does not extend into the remote countryside and areas where broad area lighting would not be economically feasible. Reflectorized lane markings, barricades, object markers, signs, and channelizing devices do help. For other hazards, the best solution might be a combination of highway design, vehicle design, and aftermarket devices. If people with disabilities have a license to drive, they may need some driver assistance devices and special vehicle design features to facilitate hazard avoidance behavior.

(n) The protective cocoon

A passenger vehicle was rear-ended by a drunk driver at a high speed. A fire quickly ensued and several of the occupants were severely burned. The police arrived, investigated, and completed the checklist options available in the collision report. They found that the cause of the accident was a driver under the influence of alcohol who was traveling at an unsafe speed. An independent investigator later determined that the placement of the fuel tanks exposed the occupants to a risk of rupture and fire at moderate impact speeds. The investigator concluded that the vehicle was not sufficiently fireworthy. Still another expert concluded that the vehicle was not crashworthy, a design defect.

Crashworthiness includes an important concept, that during a crash the space containing the occupants should remain intact (a safe space). There should not be any intrusion or crushing sufficient to injure the occupants. This requires a structural cage or protective cocoon, aided by other design features intended to absorb energy by metal deforming (crush), by distribution of the crash energy throughout the vehicle, and other energy management features designed into the vehicle.

While three different causes were found by three different specialists (drunk driver at excessive speed, a lack of fireworthiness, and that the vehicle was not crashworthy), each is correct under his own understanding, past experience, unique perspective, and familiar guidelines and standards. One blames the driver and two blame the vehicle manufacturer. It also illustrates the difficulty in counting or tabulating causal factors from just one source. The design defects can be corrected, but the driver fault cannot be completely controlled. Thus, there is an emphasis on design remedies to minimize the risk of intense fuel-fed fires, to achieve proper energy management, and to provide a safe space or protective cocoon for the vehicle occupants.

Some vehicles may not have strong A, B and C pillars and crush-resistant roof structures. For example, a convertible has an open top. To assure a protective cocoon for the convertible occupants, some form of roll bar should be incorporated into the vehicle structure. The roll bar may be actuated quickly at the start of a rollover (pop-up into position) and the pretensioner actuated to remove slack from the belt restraint system. This serves to restrain the occupant within a structural roll cage. Addition of airbags intended to maintain the restrained occupant, including arms and hands, within the structure would improve the protective cocoon. Three point belt harnesses could be replaced with four- or five-point restraints if needed.

Construction equipment is used in rough terrain under varying load conditions, and may have inherent stability problems. The likelihood of tipover and rollover is ever present. For some time, operators believed they could jump to safety as a rollover began, but rollovers can accelerate very rapidly and entrap or crush the operator. Many such vehicles are now factory-equipped with rollover protective systems (ROPs), including a safety cage, and occupant restraint belts. It is better to keep the operator within a protective cocoon than to permit him to test his ability to know when to jump free.

Likely scenarios where protective cocoons serve a useful safety purpose include the following examples. A dozer or tree dozer may push over trees, but some branches will break and fall on the operator. A front end loader may be nose-heavy with a loaded bucket and could tip over frontward. If the load is carried high, there might be over-balancing and tipover in any direction. The pitching of tracked vehicles, when walking over ridges, rocky areas, and logs, may cause a turnover backwards.

As a supplement to protective cocoons, there may be hydraulic outriggers to increase vehicle stability for cranes, truck-mounted shovels, and other mobile construction equipment.

In general, a construction equipment rollover protective system includes the protective cage structure, an occupant restraint system, a warning device to alert the operator if the occupant restraint system has not been properly fastened, possibly a warning device to signal impending rollover, and information concerning ill-advised actions such as jumping during a rollover. The seats may have side wings or bolsters to keep the occupant's upper torso upright. Where there are moving structural elements immediately adjacent to the operator's compartment, a wire screen may be necessary to separate or isolate the occupant from the moving crush or pinch point.

Falling object protection systems (FOPS) have evolved, in general, from simple concepts of brushguards, overhead protection, and canopy protection to become more integrated crush-resistant and occupant protection cab structures.

There are SAE standards for *overhead protection* for forestry equipment (J1084) and agricultural tractors (J167). There are SAE standards for *rollover protection* for industrial and agricultural tractors (SAE J334, now canceled), motor graders (SAE J396, now canceled), crawler tractors and crawler loaders (SAE J395 now canceled), wheeled agricultural tractors (J1194), rubber tired front end loaders and dozers (SAE J394, now canceled), and construction, earthmoving, and mining machines (SAE J1040c). Canceled standards may be superseded and available in a different form. The ASAE also has standards for agricultural tractors S305 and S306). Even canceled standards can be useful for design safety purposes.

For further information on construction equipment, see MacCollum (1988) and Nichols (1962) at pages 3–35, 12–77, 14–55, 15–48 and 49, and 15–69.

(o) Compliance-plus

A serious flaw in the design process occurs when trade standards are used to establish goals or objectives to be achieved. They should always be considered as minimum requirements that must be exceeded, both in depth and in scope. That is, the objective should be compliance-plus.

Standards cannot cover all the attributes that are desirable, and they are only an abbreviated version of the standards that formulators believe are necessary. Most standards are voluntary consensus documents where compromises have been made by those who wanted a higher level of safety. They may intentionally omit those provisions that have provoked conflict among the formulators because they were thought to have a high cost for compliance, involved some technical difficulty (usually test verification), or where the provisions would impose a hardship on some manufacturer already selling a particular product line. Revisions to existing standards may delete provisions that have had unintended consequences to a company represented on the standards committee.

Standards such as those carrying the imprint American National Standard (ANSI) may have been prepared, printed, and distributed by a vested-interest trade association and a selected distributor. Government standards are, of course, generally mandatory, but an added margin of safety is necessary to account for normal variations in the compliance process. The SAE has published helpful Standards, Recommended Practices, and Information Reports. They are of high value in terms of standardization of items from suppliers for mass production assemblies. The SAE standards are very useful and participation on the standards committees is instructive on perceived problems and technological challenges.

Many standards, on a wide variety of topics, have been published by the American Society for Testing and Materials (ASTM). For example, ASTM F 1337–91 is entitled *Standard Practice for Human Engineering Program Requirements for Ships and Marine Systems, Equipment, and Facilities*. It is to be used in conjunction with ASTM F 1166–88 entitled *Standard Practice for Human Engineering Design for Marine Systems, Equipment, and Facilities* (a 165-page document).

Standards widely used in the automotive industry are available from the American Welding Society, American Society for Metals, and the National Fire Protection Association. Many other standards are surprisingly relevant to automotive safety and are sometimes quite instructive. Rarely does one standard fit all needs.

> The overall regulatory standard-making process has been broken for some time ... we have also seen the lawsuits filed against ACGIH's TLV process and ... whether we like what we have seen or not, there is a new reality.
> (Henry B. Lick, PhD, President, American Industrial Hygiene Association and Ford Motor Company engineer)

Asbestos had been used in brake and clutch linings long after it was known to cause asbestosis, lung cancer, and malignant mesothelioma (Peters and Peters, 2001). The problem was a reliance on 'safe' or 'permissible' exposure standards published by both governmental and professional organizations. The acceptable limits kept being lowered as they proved to be insufficient or inaccurate. Finally, a total ban on all forms of asbestos was instituted in the European Union. But the resulting prevalence of asbestos diseases has been called epidemic in proportion to other public health problems. The health consequence and liability costs have been very high. This clearly indicates that standards are not just a goal for compliance, but a minimum that should be exceeded according to the specific circumstances and the informed risk that is acceptable. (See Chapter 3, section (b)(4), 'The compliance model'. Also see Peters and Peters (2001).)

(p) Universal design

It is only natural for an engineer to design a product that he feels personally comfortable with, can operate easily, has a good dimensional fit, and believes will be accepted by the average user. But all the users may not be of the same height, weight, age, intelligence, or have the same familiarity with such products. They vary considerably in terms of human attributes and variables. Most differ in some respect and degree from the average, mean, or median that may be in the mind of the designer. The engineer may be a college graduate with considerable mechanical aptitude, with relevant skills and knowledge, and having gained considerable understanding of the product from the learning that occurs during the design process. In other words, the design engineer is generally a poor representative of the broad range of potential purchasers or users of a product. The product should be user-friendly as well as designer-friendly.

Instead of using the average male as the design criterion in the automotive industry, this was broadened in scope to include 90%, then 95%, then 97.5% of the male population. Then females were included. Then children were included. This required designs that could accommodate or be adjustable to fit different users such as the manual adjustment of bench seats front to rear. This user-friendly approach is known as *universal design for people*.

But, to obtain designs that are actually universal, they should be capable of operating over the range of environmental conditions that can be anticipated. This would include operation at the extremes of temperature, on paved and unpaved roadways, with ice and snow on the ground, and with predictable loads and speeds. This is *environmentally friendly universal design*.

In addition to people-centered and environmentally-compatible design objectives, the concept of universal design also applies to the purposes for which the vehicle is intended as well as the unintended purposes or uses of the vehicle. It may be labeled as a five-passenger vehicle, but if it can be used for six or more people with an additional heavy load, it will be used that way. It may be used to do all things for all people. Universal design contemplates not just what is reasonable to the manufacturer's engineer, but what is foreseeable in actual use. It is *universal design for anticipated use*.

Some manufacturers sell models that can be customized with trim changes, factory options, and aftermarket accessories. Some models are made for slamming to achieve different, flashy, fast, and attention-getting vehicles. A *slam car* might have added racing wheels and tires, boosted horsepower, lowered suspension, spoilers, special body panels, high powered stereo sound systems, and special color designs. The factory might sell a preslammed vehicle, usually a cheap entry-level model, with knowledge that it will attract a special group of younger buyers. The vehicles have been sold for specialized predictable modifications. The basic vehicle design should be compatible with such modified or altered use. This is *universal design for predictable modification*. It is anticipatory design safety. Universal design does impose a significant burden on the design engineer, but reduces the burdens on the purchaser, drivers, and on society in general.

For additional information on this topic, see Chapter 6.

(q) System engineering

System engineering deals with the interaction between two or more parts, components, assemblies, or products. Each of the elements of a system may have been designed and produced independently, but when combined they may affect each other and the system in some unexpected manner. The interaction effects may not be revealed by formal design reviews, prototype or preproduction testing, and may be discovered rather late in the design and development process. They may be discovered only after a product is procured and used by another firm or customer.

For example, a recall of 12,000 vehicles occurred after passenger airbags were actuated without a collision. A buildup of static electricity could result from starting the engine, closing a door, or manually wiping the instrument panel. The remedy was a dealer-installed grounding strap. The problem was a result of interactions between the passenger airbag system, the accumulation of static electricity from several sources, cool dry climates, and lack of sufficient grounding.

The interaction problems arise because design often takes place one component at a time by somewhat independent persons or groups. For example, the brake engineers may not be able to talk openly with the suspension engineers, or the windshield wiper vendor may not have his voice heard by the assembler even though reliability and interference problems become known. Early and open communication may be difficult, but it is a necessary preventive measure, particularly for safety-critical functions.

Another system engineering problem is in the design, selection, and installation of electronic systems for vehicle telecommunications and other electronic devices. Unwanted system interactions can occur from the electromagnetic compatibility (EMC) problems of wires, cables, and connectors. There may be crosstalk and radio frequency (RF) emissions from inductive, capacitive, and antenna effects depending on the contact materials, continuity, surroundings, operating frequencies, and leakage from shields. The system designer may be alert to the complications of bundling wires, the need to segregate wires to prevent crosstalk, avoidance of highly sensitive circuits, the need to reduce circuit bandwidth, the benefits of short wire length and coupled or twisted-pair circuit loops, the effects of hot insertion and removal, and the need for adequate shielding of problem circuits. This should be accomplished early in the system design. Particular attention should be given to connectors that reduce the possibility of installer error.

Marketing is king because marketing personnel remind others that they produce the sales that pay salaries, overhead, profit, and stock dividends. Until recently, safety had a comparatively weak or disregarded voice in terms of late design changes. Therefore, marketing may have a very strong voice in terms of late changes, adding devices, or requiring some brand features that could command the attention of prospective purchasers. These late additions may produce unexpected interaction effects that could adversely affect both system performance and customer relations. If there are late changes, the interactions may be discovered only after the vehicle is delivered to the customer.

A specific system engineering effort may be necessary to maintain a broad perspective and ferret out interactions that may be harmful. The use of computer-aided design serves to open-up the design of each element to inquiries about interactive effects, but some form of early testing may be necessary to assure that the unexpected does not happen or is corrected if it does. In some organizations activities of reliability engineering (failure mode and effect analysis) and system safety engineering (fault tree analysis) may serve as part of a system engineering effort. Of course, human factors engineering is primarily devoted to treating the interactive effects between man, machine, and the use environment. Finally, coordination between design groups should be facilitated, not hindered by organizational walls that prevent open communication.

An illustration of the human–vehicle interface is the entry into a heavy truck or large construction vehicle. The old safety adage is that there should be three points of contact at all times (such as two hands and a foot). Animation may be used to show the reach envelope for the right hand of the 5-percentile female to 95-percentile male. The human subject may reach for grab handles, the steering wheel, or other aids to climbing up and into the vehicle. The effects on muscle groups and bones may suggest that injuries could be avoided by better placement of grab handles and steps.

(r) Survivability and injury reduction

(1) Contagion

There should be no question but that a design safety specialist must hold uppermost in his thoughts and actions the basic concept that accidents can be made more survivable and with less injury. This belief can be contagious if means are developed to communicate with other engineers and reinforce their moral and ethical duty to help save lives and prevent injury. Contagion is a necessary preliminary to team efforts to reach high standards of health and safety. Spreading the virus of safety should be accomplished in a pleasant, easily acceptable, and informative manner to avoid challenging peers and embarrassing coworkers. But the goal of increased survivability and injury reduction will not be achieved without steady reinforcement of the concept that this is an important design requirement that should not be overlooked or neglected.

(2) Anticipating the possible

There is a car culture of hotrodding, drag racing, and conversion to racing machines. So when one 1999 model passenger vehicle was advertised as being capable of a rated speed of 115 mph, should it surprise anyone that the vehicle was driven by purchasers at high speeds and privately tested on the road to see if it could reach the 115 mph rated speed? One driver proudly stated that he had exceeded the speed by nearly 5 mph. The advertising established a goal and the high price of the automobile created an incentive to prove that the rated speed was real.

Driver behavior can be predictable under certain circumstances. The operating hypothesis is that if it can be done, someone will attempt to do it. The design safety engineer should anticipate the possible. This may be called a worst case of foreseeable human behavior, but it helps to establish a design requirement for survivability. If the vehicle goes out of control at a high rated speed, is there sufficient crashworthiness or an electronic vehicle stability system capable of regaining directional control? Speed governors and recording devices in commercial trucks are attempts to control speeds by downgrading the possible.

Similarly, if a sports utility vehicle (SUV) is advertised as having an off-road capability, someone will use it for off-road travel. But how rough the terrain, how adverse the weather and how wet the ground, and can it climb over the rocks and hills shown in television commercials? Everything may be a matter of degree and reasonableness, but what expectations and goals are created for the driver? What are the typical driver response and creative attempts to prove the utility of the off-road designation? It is not too difficult to envision and anticipate the possible situations in which the driver might operate the vehicle. The ultimate questions are whether the vehicle has actually been designed for such operations and exactly when the limits might be exceeded?

Anticipating the possible is an intelligent and knowledgeable form of prediction. It is the kind of foreseeability that should be key in design safety engineering and the marketing of automotive vehicles.

(s) Digital models and man-testing

A mechanical device can be tested to its extreme physical limits and any failure should provide useful design information. The failed device can be easily replaced for further testing. This can be done with vehicle components, assemblies, and the entire vehicle. But place a human in the driver's seat and extreme caution must be exercised to avoid injury. Yet the combination of the vehicle plus the human driver is what may be needed for realistic testing. The human is generally the limiting factor in automotive vehicle design and its most difficult and ambiguous safety consideration.

To avoid injury to human test subjects, available data has been accumulated, extrapolated, and used to formulate mathematical models. Tests then can be run and data used in the mathematical models to achieve a desired approximation, result, comparison, or indication of compliance with a specification or standard. This could be supplemented by human design criteria such as anthropometric dimensions for age, sex, ethnicity, and nationality. Disability, sex, morbidity, and mortality data is available. In addition, information is available as to the mean and variance of human reach, grasp, fit, and movement. Limited relevant behavioral data can also be used.

For better design visualization of the interface and interaction between vehicles and humans, three-dimensional manikins have been utilized effectively. For crash testing, instrumented dummies have provided useful data, were expendable, and meant that humans were not harmed.

Motion picture animation has become more lifelike, morphing software has become available, and human animation-based simulations have become increasingly useful. Evaluation studies of devices such as power adjustable brake and throttle pedals could be accomplished very quickly and the graphics were immediately comprehensive and convincing. The applications of human animation became useful in workplace layout, material handling, and machine tool operation.

Human simulation is an effective design tool. It is compatible with computer-aided design and manufacture programs. Physical mockups and prototypes should still be used, although the temptation is to shorten the design cycle by eliminating as many steps in the design process as possible. However, human-induced problems require appropriate study and the digital model is a quick and effective tool. The basic issue is whether there has been adequate evaluation to assure that the vehicle has been appropriately man-tested for all foreseeable uses.

For further information, see Chapter 9 and Chaffin (2001).

(t) Design by and for test

At a vehicle assembly factory, subassemblies were being delivered by truck on a just-in-time basis. This meant that there always would be a 4-hour supply on the floor adjacent to the location where they would be installed on assembly line vehicles. The partially completed vehicles were mounted on a dolly that was moved by a ground level conveyor system. There was no subassembly pre-installation test, and no incoming quality inspection, so great reliance was placed on the first-tier vendor's assurance as to quality and function. Throughout the vehicle assembly line there was very little quality inspection or testing. When completed, each vehicle was tested for basic operation. That is, could the engine be started, the tires rotated on supporting drums, the instrument panel displays and controls operated, and the vehicle appearance checked? If there were a problem, such as a paint scratch, the vehicle went to a large repair area for individual attention. In essence, most of the quality control and testing was delegated to the suppliers and, later, to the dealers or make-ready facility.

To illustrate one potential assembly problem, the main wiring harnesses were received as a tangled group in a large cardboard box from the supplier. One would be fished out, hung on a hook to untangle, and then thrown into the engine compartment. The instrument panel, with attached harness, was installed and the harness fed through holes in the firewall into the engine area. The feed-through of the harness branches was a manual operation, a push and pull on a feeder attachment. There was no test of the wire harnesses, and the only test when the vehicle was completed was whether the connected equipment would operate. This crude test would not uncover damaged wire and connectors, something that is of growing importance with the sophisticated electronic equipment now being installed, plus the requirements for extended durability, compensation for age deterioration, and avoidance of warranty claims.

Building-in test points, for both assembly testing and dealer troubleshooting, could include quick testing for more parameters than simple operation. For example, a customer might tell a dealer's service representative about intermittent problems with the vehicle's electrical or electronic systems. A warning indicator might go on, perhaps indicating a light bulb failure in the turn signals, but slamming the door might cancel the annoying warning indicator. Is there a warning circuit problem or a problem with the turn signals?

In the past, large modules were replaced if there were any signs of malfunction in order to simplify the service diagnostics. But unhappy customers had to shoulder the cost burden. It was favored by dealers as one means to run a very profitable service and parts replacement operation. Test points permit pinpointing the actual cause of the trouble and indicating the least expensive adjustment, repair, or parts replacement.

Recent extended warranties shift costs to the dealer and the factory, so greater attention is being paid to cost savings. The future suggests a greater variety of vehicle electronic components and systems, some networking and some interacting, some single function and some multifunctional chips, all of which could benefit from more pin-pointed built-in diagnostic analyses and repair procedures. For this design requirement, properly designed diagnostic software and accessible built-in connections for test equipment could help the dealer, the customer, and the factory. In essence, the vehicle should be designed for testing rather than attempting to find a means of testing the vehicle.

There is one caution: the design of test equipment should not be an afterthought, since it could harm the vehicle components. Even the test equipment should be thoroughly checked, since past experience has indicated that electrical, mechanical, and fluid problems might be created that could damage the test equipment. That is, there should be a means to test the test equipment.

(u) Design for manufacturing and assembly

What might seem to be a minor design engineering change could cause a major increase in manufacturing costs. An accumulation of higher fabrication costs might make the product noncompetitive in the marketplace in terms of sales price, and adversely affect company profits and stockholder dividends. Tight control of manufacturing costs is essential for corporate health and survivability.

A disciplined approach to manufacturing (budgeting) cost control is needed as warranties and guarantees are extended in time and coverage. The product must continue to function; to meet customer expectations from the extended warranties, for a longer period of time, or the dealer and assembler must pay the price. The necessary durability and quality must be built-in by manufacturing and assured by engineering actions.

It is the basic product design that determines the capital expenditures needed for new dies and machine tools, the production cost per unit, and the eventual warranty costs. This is why many companies have some form of liaison or coordination between engineering and manufacturing, to reduce departmental isolation and improve communication. In an early design review session, a manufacturing representative may be able to suggest changes to enable the use of existing production machinery and processes. He may ask what gain may result from a design change, in contrast to the increase in manufacturing cost. He may wish to use an alternate part or subassembly from already qualified suppliers with good records for timely delivery, consistent price, high quality, and conformity to specifications. He should be able to estimate labor costs and contact vendors for comparative prices. The manufacturing representative may be able to indicate how the product might be inspected, assuming engineering indicates what attributes and variables are desirable, what is actually important for proper function, and what is critical to safety.

The manufacturing representative may be able to estimate, based on the skills and tools available, the expected workmanship defects and how they may affect the product. The defect rate may require special controls or may serve as a reason to outsource the item.

There is always the danger that some manufacturing specialists will unilaterally make substantive design changes. In one example, a product failed catastrophically from an early manufacturing change made to enable a subassembly to be inserted (installed) more easily (a better dimensional fit).

The manufacturing engineer should question whether the tolerances (permissible variation of specified dimensions) shown on the engineering drawings are economically and technically justified. Is the high-grade and expensive type of material really necessary? Could the bending, shearing, drawing, forging, and other fabrication operations be simplified? Has a value analysis been performed? Can there be interchangeability of parts and vendors? There are many questions that need to be asked for effective communication and the resolution of problems that could arise at a later date.

Efforts at cost savings continue throughout the production of a product. An open invitation to change is the term 'or equivalent' in a specification or bill of material. This is permission for a design change without the need for further approval. Who is to judge what is actually equivalent, particularly when the part or material could result in cost savings and gain personal credit to a manufacturing or industrial engineer for the savings? For example, during acceptance testing, some seat belts failed certain prescribed company tests, but were nevertheless utilized because they seemed to meet government standards. Some belt material failed to have antioxidant and antiozonant additives, but met the specified company tests. Both seat belts deteriorated over time and their failure resulted in enhanced injuries.

Design for manufacture and assembly should be concerned with quality inspection, since the assembly line supervisors are under unceasing pressure to maintain economical production rates and 'get it shipped'. Defects may be ignored or waived by engineering.

Unavailable materials and parts may result in substitutes being added to maintain production dates. Engineering should maintain some oversight to prevent product degradation in the actual manufacturing and assembly processes.

Engineering should look at the repair area, at the end of the vehicle assembly line, to identify problems being created by design, fabrication, and procurement. Such problems might be remedied immediately or be a lesson learned for the future. Another area of concern might be the fulfillment of special or custom orders that require a higher level of worker skills and logistic organization. The question is always: has the design been compromised and potential safety problems created?

At an early stage, there should be an agreement between engineering and manufacturing to limit design changes; be it in number, substance, or timing. Late design changes can be very disruptive to schedules and have exorbitant cost consequences.

In essence, safety problems can arise during manufacturing operations unless there is a conscious attempt to design for ease of manufacture and simplified assembly. The continuous emphasis on cost savings could erode the safety objectives if not properly controlled and audited. Design responsibilities for manufacturing and assembly start early and continue during the manufacture of the product in order to assure appropriate compliance with the design specifications and design objectives.

(v) Design for maintenance, repair, recycling, and disposal

In the past, the after-sale considerations by the manufacturer were minimal. The design engineer might not know who purchased the product, where it went, how it was used, or whether the customer was satisfied with it. The out-of-sight, out-of-mind adage prevailed. Logically it seemed that it was up to the dealer or retail seller to take care of any field or customer problems. The aftermarket accessories and modifications also were believed to be someone else's obligation and responsibility. If there were a customer relations department, it generally remained isolated from the engineering design department. In some industries, this was counteracted by a strong effort to design for maintainability. But once the product left the shop, almost all the design engineering effort was focused on the next new product. One solution was to develop a maintenance demonstration facility, using people representative of those who would perform maintenance, service, and repair operations. This was accomplished with prototypes, so that there was still time in the design cycle to correct any deficiencies, and usually there were many. In other words, the design engineers understood that their designs would be evaluated and the results documented, so concepts relating to maintainability had to be considered in detail at an early stage. Animated computer models and physical mockups could be used with design flexibility even earlier in the design cycle. A search for lessons learned in the past could also reduce repeat problems in design.

The design-for-maintainability effort was to simplify and make easy all of the anticipated tasks, but a prime focus was to identify errors and near-errors so that there could be early corrective action. The draft service manuals, technical orders, and troubleshooting instructions could be corrected as errors or needs were found. Since most mechanics avoid using service or repair manuals, the effects of non-use should be studied. Their readability and comprehension levels should be evaluated. This should include all support equipment, even that of competitors or aftermarket suppliers, if that equipment might be used on or in conjunction with the product in some situation. The object is to avoid damage, by anticipatory design features, from the use of unauthorized accessory equipment.

Reinforced plastics and carbon-based (carbon fiber) composite materials may have desirable design properties in terms of weight savings, strength, and corrosion resistance. They may perform well in factory tests, but some have experienced field problems. For example, one automobile bumper tended to explode (brittle failure) on impact, rather than progressively yielding (absorbing crash energy) as does a steel bumper. In terms of repair, automobile mechanics may have difficulty dealing with the unfamiliar materials in a part, panel, or structure. For inspection, it may be difficult to see cracks in plastic members. In other words, standard inspection and repair procedures may not be appropriate for plastics. The design engineer may have to balance the benefits of combining several metal parts into one plastic unit in order to make vehicle assembly easier, as balanced or compared with any maintenance, inspection, and repair difficulties created for the automobile service and repair facility.

The make-ready, installation, and preparation for reuse (preowned, used, or second-hand equipment) operations should be evaluated.

Another neglected area is recycling and final disposal. It is considered environmentally friendly to utilize materials that can be recycled and, in some countries, there may be government regulations to encourage it. Final disposal is a design consideration because there are always service life considerations; that is, how long should this part function before it malfunctions or fails? What is the part reliability and how does it affect overall system reliability? How long will this plastic material last before it shows signs of deterioration, appearance flaws, or wear-out? What parts may need replacement, at given times, and should notice be provided in scheduled maintenance instructions and manuals? What is the relationship between part life and overall life? Should there be planned obsolescence, under some customary customer purchasing practices or schedules, or marketing objectives for new replacement models or product upgrading objectives? For automobiles, there may be several service life design objectives within the engineering department of one company. For industrial equipment, the operating concept may be that the product can last indefinitely with parts replacement and proper maintenance. In some countries, there may be government-mandated manufacturer and distributor buy-back obligations for refurbishing, rebuilding, remanufacture, recycling, and ultimate disposal.

In the automotive industry, a commonly used vehicle service life is 8 to 10 years. This is the life span of the vehicle in terms of the original design requirements, a 'design to' objective. This does not anticipate accelerated or premature aging, mechanical wear, or environmental degradation. In other words, vehicles over 10 years old are beyond their life expectancy. Some vehicles on the road have outlived their life span.

All of this pertains to the cradle-to-grave or concept-to-death rule in product design. That is, those who design and sell products cannot close their eyes to the follow-on consequences. They have a responsibility or company and societal obligation to consider and design-out downstream problems that could result in personal injury, property damage, or environmental degradation.

(w) Appropriate marketing and leasing

Advertising representations establish consumer expectations. If the expectations are exceeded, there is a happy customer and future sales. If expectations are violated, an unhappy customer results. It is inappropriate marketing to engender expectations that cannot be met and this might create safety problems.

A marketing plan might be for each brand to have one safety feature that can be promoted. This may distinguish each brand, but the consumer may believe that the manufacturer has such features in each of its brand models. For example, a luxury vehicle salesperson was asked about stability control systems for one model and he responded, 'they all have them'. Actually, only one model on the showroom floor had such a system as standard equipment, for a few others it was optional for extra money, and for others it was not available. The salesperson created expectations about the presence of a safety system, but to others this might be considered just 'puffing' (exaggeration) or an attempt to 'control the customer' and avoid negatives in order to 'close the deal'.

Another long-time customer questioned the warranty and service, having had some minor problems. The salesman was quick to answer: 'You are right, but we have corrected that situation'. An expectation was created that the service could be relied upon and warranty work would be honored.

One very profitable automobile leasing company targeted certain professional groups as customers. It selected one vehicle brand for its fleet and gave its customers firm assurance that it would stand behind the vehicles and properly service them. An engine (motor) problem developed, but the manufacturer refused to acknowledge that any such problem existed. The lessor attempted to pacify and please the lessees because they were a good customer base and he believed that the marketing promises had to be satisfied to retain business goodwill. Vehicles returned by the customer or just left on the street could not be leased again because there was no remedy other than replacement with another troubled engine. Within a short time, the lessor's customers deserted him and he had to file for bankruptcy. In essence, leasing is primarily an independent marketing operation and there is peril in a one-brand operation and steadfast representations as to vehicle quality, performance, service, and safety to the customers.

Airport car rentals illustrate another safety problem: the lack of universality among passenger vehicles. Anyone with a credit card and a driver's license can obtain a vehicle, day or night. A driver may arrive at night, tired from a long journey on an aircraft, perhaps from another country, and receive any model vehicle that is ready and available. Familiarization with the vehicle may be in the dark or as the vehicle is being driven. If gasoline is required, the fuel cap cover lock may be released by a lever under the dashboard, on the door, under the seat, or there may be no lock. As he drives, the control locations and road-feel may be quite different from what his custom and habit dictates. In other words, the transition to an unfamiliar vehicle may create problems because of a lack of universal design for the human interface.

The general public are well aware of their right to drive a vehicle. This means all persons and all vehicles. It is a tribute to the automobile industry that vehicles are considered such a vital necessity for everyone everywhere. But the right to drive includes those with physical and mental disabilities, those who may consume drugs and alcohol, those who become fatigued and have poor judgment, those who have personality disorders and suffer road rage, and those who have bad driving records. This is a challenge to engineering specialists and should be a deep concern for those in marketing and leasing.

> Customers are a good thing, by and large, provided they're kept well downwind
> . . . customers don't know what they want. They never have. They never will.
> (Stephen Brown, *Harvard Business Review*, October 2001, p. 83)

The emphasis on determining customer needs and producing products that meet those needs has been vigorously criticized. There seems to be an undercurrent of resistance that reflects an early automobile maker's remark that the customer can have any color he wants as long as it is black. The automobile salesman wants to sell the vehicle on the floor, not one he might be able to find elsewhere or have to order from the factory. Customer-centricity has been labeled pandering to customers (see Professor Stephen Brown's comments above). This may not reflect what some vehicle manufacturers really want. They believe that they need a more efficient supply-chain management system to reduce inventory supply needs, to improve their demand forecasts, and to reduce operating costs per vehicle. One attempt to accomplish this is the Web service for a build-to-order manufacturing and distribution system for vehicles on a short lead time basis, with a broader range of options, and retention of the independent dealer system including megastores.

(x) Recall and liability avoidance

Every company should have a formal recall program plan. It should state who has what responsibility, how a recall could be initiated, and the means by which it should be implemented. If it is to recall for an unsafe condition, it should be done quickly and reach all of the owners or users of all products being recalled. Any confusion and delay might result in injuries, damage, mandatory penalties, or adverse public relations.

Advance planning is necessary to determine the proper form and content of notifications, the most economical approach consistent with actual notice to the ultimate user, and any required governmental agency notices. There are two typical problems to be confronted. First, direct notices to the consumer are notoriously inefficient, with compliance rates of 60% to 80% for some products even with repeated notifications. Multiple methods of notification may be required. Second, costs can quickly escalate if too many company employees get involved, high-end delivery rates are used, and available means within company control are disregarded. For example, the Internet can be utilized, the service departments of dealers can be alerted and provide information to screen vehicles, trade publications may carry free or low cost notices, a 1–800 telephone number could service requests for information, and marketing could make personal contact with those within the distribution network. Return cards completed after purchase, warranty files, serial number cards for product history, and dealer or distributor customer lists may be used to identify and locate those who should be notified. Computer printouts of addresses and notices drastically reduce recall costs and provide evidence that a notice was, in fact, sent. Some companies prefer notification lists to be maintained by an independent company so that this fact cannot be questioned at a later date.

Much depends on whether the recall is voluntary or mandatory, quiet (secret) or loud (to gain derivative publicity), claimed to be a free product enhancement for the company's good customers, or a very serious safety, litigation, or regulatory agency problem. A recall has been successfully used to provide a minor benefit to the customer, but has been accompanied by advertisements of new accessories and new products to ensure that the recall was profitable.

A recall plan must be periodically exercised, so that the participants become familiar with their role and responsibilities, the activities take place quickly and harmoniously, and any necessary updates can be included in a revised recall plan. Since the notifications may be disregarded if a safety problem is masked or rendered low-priority, a focus group could be used to determine whether it invites compliance or may be misinterpreted and disregarded. Of course, any detailed study should be accomplished as part of the exercise or simulation of the recall plan, so that any actual recall would not be delayed.

Liability mitigation is the attempt to show good faith or absence of fault for an unsafe condition or defective product. A timely recall program is some evidence of good faith and proper intentions. It should appear to be an attempt to protect the customer and provide a remedy just as soon as the company realizes that it is indeed a safety problem. Thus, the company acts as a good citizen. Documented design safety efforts provide evidence that conscientious, costly, and substantive efforts were made to identify and correct safety problems before the products were sold to the purchasers or used by other people. These efforts might include a formal design drawing approval system, formal design reviews, various test programs, safety devices, safety features, focus groups, claims comparisons, and product history analyses. Liability mitigation is not prevention, it is a means to reduce the penalties or percentage of fault. Liability remains as a motivating fear and prompter for safety assurance efforts.

(y) The informed purchaser and operator

In the future, the automobile dealer may be required to provide assisted learning and much greater help in the decision-making sales process. The 'informed purchase principle' refers to two aspects of the sale: first, that the driver fully understands what is being purchased; second, that there is meaningful instruction on risk avoidance behavior for the particular vehicle being purchased.

In the past, it was something of a dealer joke to ask a potential female purchaser only one question: 'What color do you like?' Requests for information were often answered with brochures that had attractive color pictures of various models. The automobile salesman rarely looked at the dealer's more detailed technical information supplied by the manufacturer or manufacturer's representative. Instead, he often pressed the customer on price, discounts, financing, accessories, and delivery time in order to 'capture' the purchaser and to 'close the sale' as quickly as possible. A quick test drive might be the only familiarization with the vehicle. Videos, if available, took too much time. After-sale instructions were considered non-productive except as a friendly goodwill gesture. When asked a specific technical question about a vehicle, a salesman might not know but would say something to please the prospective buyer. This type of paucity of information scenario might be appropriate if all vehicles were more or less the same. After all, during the past 100 years of automobile use, every driver has had ample opportunity to become familiar with the basics and most vehicles have had only minor styling differences.

Now vehicles are becoming far more complex and different from others. The road or traffic mix has become a lesson on diversity and uniqueness. Some vehicles may be high-speed, high-acceleration, and high-horsepower vehicles with off-road capability and ample towing capacity. Suspension, load-carrying capacity, and steering responses vary. They may be loaded with electronic devices such as multiple television monitors, communication and entertainment devices, and navigators. There may be a conversion to a wired car with e-mail, fax, personal computers, video games, snooze alarms, and carry-on accessories. They may be very small or very large vehicles, some have side panels or steering systems effected by wind gusts, and some can be loaded in an unstable manner or effect oversteering or understeering. Some may have electronic stability systems, others do not. Each vehicle may have different safety characteristics for a particular customer use.

Only the dealer has a direct personal contact with the buyer. The dealer's salesperson sees, talks, assesses, and interacts with the buyer. Is there any dealer responsibility, other than just the sale? The dealer may have an obligation to recommend a particular model that has a better 'fit' for the driver and meets the 'needs' in an appropriate fashion. The salesperson may see the fit and spot possible safety problems during a test drive of the vehicle.

The following examples may be instructive. A taller than average driver visited a retail dealer showroom to look at a luxury import passenger vehicle. He wanted comfort for his supplemental office-on-wheels and the interior of the vehicle looked spacious. After he purchased the vehicle, he noticed that the most comfortable position was with the seat fully moved to the rear. His knees straddled the steering wheel and rested on the padded instrument panel. However, this position was somewhat awkward for him to keep one foot on the accelerator pedal and the other poised to depress the brake pedal. It is assumed that the dealer's salesperson was aware of the poor fit, but did not want to risk the loss of a quick sale. In another situation, the driver's head was close to the roof, but there was no problem until the seat was adjusted forward. Then, his head would hit the sun visor or the roof if he moved his head forward, attempted to look to the side, or twisted his head to talk to other people in the vehicle. In still another situation, a large blind spot to the right rear was discovered when a lane change was initiated and quickly aborted because of the presence of a vehicle in the blind spot. In still another vehicle, the best position of the driver resulted in her eyes being just above the steering wheel. One driver selected an SUV to provide ample room for his overweight condition, but could not reach all the instrument panel controls without contorting his posture. In each situation, the salesperson could have recommended alternatives to the accommodation difficulties.

Only the dealer is in a position to determine whether the driver's seat is moved too close or too distant from the airbag location on the steering wheel; such as the 10 inch (25.4 cm) distance criterion. The driver's head may be positioned more than 1 inch (2.54 cm) from the head restraint or the top of the driver's head may be above the top of the head restraint (head rest). The dealer may determine the need for brake and accelerator pedal extensions or the proper settings if the extensions were built into the vehicle. The dealer may change the seat anchorage location for oversized drivers. Thus, the dealer can take some action about possible safety problems and attempt to make the buyer an informed purchaser and operator of the vehicle.

There is a difference between 'tailored vehicles', based on the unique customer needs as determined by the dealer, and 'customized vehicles' which are standard selections, options, special paint jobs, trim changes, and available accessories. The dealer would seem to have greater responsibility for the tailored vehicles since the configuration is accomplished by the dealer. The manufacturer may have greater responsibility for customized vehicles, particularly if they are special-order factory configurations.

Some vehicle manufacturers have developed *pre-delivery inspection procedures* for each vehicle. This may include a road test for squeaks, rattles, and wind noise; a check of the horn, displays, switches, radio, turn signals, windshield wipers, clock, fluid levels, lights, etc.; and a check of appearance items such as protective cover removal, floor mats, trim, and finish. This check list is to be certified by the service technician and service manager. There may be a *customer delivery procedure* in which the customer agrees with the salesperson that a pre-delivery inspection was made, and that the salesperson familiarized the customer with features described in the owner's manual including the seatbelt system. In essence, it serves to shift responsibility to the dealer for the inspection items and also to the purchaser who acknowledges in writing that everything has been explained to him. This is *not* what is meant by an informed purchaser, since it is rather cursory, often ignored, and does not focus on safety.

In summary, the dealer is in the best position to make adjustments in the vehicle (such as head restraint height) and to explain to the driver why such adjustment is necessary. He can advise the customer about the benefits of seatbelt pretensions and caution about out-of-position problems. The dealer can reinforce the warnings in the owner's manual that indicate that reclining seatbacks should always be in the upright position while the vehicle is in motion. As new vehicles appear on the market with different steering characteristics, hands-off automated operation (intelligent vehicles), and other systems unfamiliar to the driver, the need for the dealer to instruct and inform the consumer is increased. Perhaps, the risk reduction (health promotion) instructional efforts by the dealer could include a combination of standard videotapes, standardized Internet information packages, supplemental factory literature and useability test results, the involvement of community-based complementary information sources, local training facilities, and salespersons who can actually communicate technical information. The informed purchaser would present less risk for other drivers sharing the roadways, as well as reducing the dealer and factory loss potential.

(z) Participatory intervention

Problems and difficulties that may have safety implications are often known to workers, but those workers may have little immediate credibility for management or have no voice in the actual operation of their workplace and the factory work system. The problems may continue unabated and unknown to others for a long time. Similarly, problems may be known only to dealer repair mechanics and to the customers who own and operate the vehicles. The problems may not be known to the engineers who could investigate and make changes. When asked about a parts failure well known to customers, a sales manager replied: 'We sell a lot of parts'. It is true that parts replacement is a high-profit operation, but it encourages counterfeit parts that loosely replicate a part using cheaper materials, cheap labor, low overhead cost, and packaging that imitates an established brand. This aftermarket for false parts is huge and tempting. But if the windshield glass does not shatter correctly or if a brake pad is made out of steel wool, there can be serious safety problems. In essence, how are safety problems discovered in the workplace, auto repair shop, owner operation of the vehicle, and aftermarket sales? There may be a need to create a reliable means to intervene in the organizational structures and functions, from design to disposal, in order to reach people with hidden knowledge, to facilitate discovery by joint participation, and to identify specific improvements in products, processes, and service.

There should be a recognition of the fact that not all people are safety advocates willing or able to overcome resistance to change that will permit incursions into their jurisdiction and power base, or are willing to foster communication likely to identify problem areas.

Any such effort should be conceived as a short duration and temporary foray into operations, conducted in a mutually acceptable fashion, with clear limits and objectives, and with a focus on problem-solving. For example, one set of participants can be facilitators and change agents, the other set of equal participants may be the knowledge holders, and the forum may be small focus groups or informal on-site individual discussions. The initial incursion may be by ergonomics specialists (see *Muscoloskeletal Disorders and the Workplace*, 2001) or others with easily recognizable primary objectives. Such an effort can empower workers and provide for human dignity or pride by overcoming bureaucratic and administrative barriers. It can help to create a general climate of safety improvement for the company.

In some respects, this is a form of empirical research or field experimentation to identify and solve generic problems. It may deal with worker biodynamics in fabrication tasks or operations, manual moving and lifting loads, better tools and equipment, and discovering the cause of discrepancies, nonconformities, rejects, and defects. It may involve mitigating work-related violence, intentional creation of defects for various reasons, and assembly line omissions. It could involve customer service problems and how to achieve higher levels of customer satisfaction. It is essentially a human-centered, quality of life, and proactive approach to safety. Unfortunately, the safety problems must be discovered in a context that deals with sensitive social interactions in carefully arranged informal relationships.

Participatory intervention is particularly desirable for knowledge-based work, that which is dependent upon technical and electronic communication, and that which undergoes constant change.

> Human relationships are beginning to depend on technology and electronic communications . . . many relationships have been negatively affected by textual misunderstanding in the electronic communication environment.
>
> (Hal Hendrick, PhD, Former President of the
> International Ergonomics Society
> (see Hendrick and Kleiner 2001, p. 105))

The rule is that defects can be inserted into the product, anywhere from design through repair, incidental to any work task or intentional act, but discoverable only by conscious directed effort. Fundamentally, a participatory intervention is an organized attempt to dig out the hidden safety problems before they can cause harm.

More than 300 languages are spoken in the United States. The linguistic proficiency for the spoken word is far greater than for the written word. The first question is to what extent do foreign-language readers and near-illiterate persons actually understand the English-language automobile sales contracts and lending agreements? The second question is to what extent do they understand automobile owner's manuals, repair shop manuals, and factory assembly drawings? The third question is whether culturally diverse populations interpret English-only documents in a context or manner that could produce operator errors or unacceptable deviations in work practices. The potential for safety problems seems obvious. The potential for participatory intervention techniques also seems obvious. In other words, find the problems and solve them before they become costly.

In a recent example of how top management can be isolated or protected from knowledge concerning serious acute injury problems, the chief executive officer of a major corporation confessed that he first learned about the problem watching a television news broadcast investigative report. The problem had been well known by virtue of numerous lawsuits against his company. The problem was not something with a sudden onset; the injuries had been accumulating over several years. Because of design changes being made on new models, the inference was that some people in the company did have knowledge of the problem. The question remains: was this sufficient for management oversight and coordination of a problem that could have major financial ramifications for the company? More basically, could a participatory intervention program have discovered the problem much earlier, communicated and coordinated it with all responsible individuals, and saved lives and money in the process?

> The Amsterdam Treaty commits the EU to a high level of consumer protection and to integrating consumer policy into other common policies and activities. So my Commission has made it a top priority to improve the quality of life for Europe's citizens.
>
> (Dr Romano Prodi, President, European Commission, *Quality of Life: A Priority for Europe*, Third Annual Assembly of Consumer Associations)

3   Risk evaluation: is it unsafe?

(a) The basic trilogy

(1) Definitions

The terms 'hazard', 'risk', and 'danger' constitute the three stages in one fundamental perspective that is repeatedly utilized in design safety. The hazard is a condition or situation that has the potential to cause harm. The risk is the character and magnitude of the harm that could occur. The danger is the characterization of the risk as being safe or unsafe. In other words, a particular hazard may manifest such a level of risk as to be considered dangerous.

Hazards may include human errors, failure modes, features, conditions, characteristics, attributes, or structures that conceivably could cause personal injury, property damage, environmental insult, or loss of the desired performance of products, components, assemblies, processes, or systems. A person may be exposed to a hazard with or without harm.

Risks may be the likelihood, magnitude, frequency and severity, extent, or amount of harm that could occur. Risk includes the character (type) of harm and its severity (one high-intensity harm or lesser multiple widespread harm). It may be a qualitative, quantitative, or mixed data assessment depending on the data available and the data uncertainties that exist. The process of determining the risk is called a risk assessment or risk evaluation. A high risk hazard, that might cause serious injury or death, generally requires a further categorization by a decision-making process.

Danger is determined by comparing the hazard's risk to some criterion, standard, or reference point. An acceptable risk is not a danger. An unacceptable risk constitutes a danger that may be considered a defect, fault, unsafe condition, safety deficiency, dangerous situation, an unacceptable risk process, a critical failure, a serious threat to life and limb, or simply a danger. A danger should be prevented or corrected, before further human or system exposure, to attenuate and control the risk or exposure.

(2) Examples

For example, a hot surface has the potential to cause burns, but if it is adequately covered or shielded it does not have a high risk of exposure and harm, so it probably would not be considered a danger. An unshielded power takeoff on a tractor could ensnare clothing, twist it and wind it on the rotating shaft, and cause serious injury to a person. The hazard is well known. If it is not properly guarded, then historical experience suggests it poses a substantial risk of serious injury, so it constitutes an unacceptable risk or danger. The alternative design, to lower the risk substantially, could include an absence of projections (bolts and angles) that could catch and hold clothing or other materials, an improved shield, and appropriate warnings. A sharp edge can cause harm by cutting the flesh of a maintenance specialist, but the risk is low if the sharp edge is covered, bent back on itself, or made into a dull edge. In a design-modified form, the sharp edge probably would not be classified as a danger, assuming a more detailed analysis was performed to assure that the edge could not cause some other harm.

The delayed failure problem is illustrated by a tire recall program. The hazard was a belt-leaving-belt tire failure mode (belt separation), the risk was loss of vehicle control ending in rollover or collision (a threat of injury and property damage), and the danger decision was that it was a safety-related defect. It was a time-delayed failure because the tread separation might occur only after several years of operation (EA 00–023, 2001). More specifically, the hazard was a belt edge crack that could propagate laterally and circumferentially into a major separation and loss of the belt-to-belt adhesion. The cause of the hazard was a deficiency in the wedge rubber, at the shoulder where the round carcass meets the flat tread, an area susceptible to heat buildup and high mechanical stresses. Even more specifically, the hazard was created by a steel belt edge that was cut during manufacture leaving unprotected steel cord edges that were sharp and could fray. The process of further defining a hazard leads to what should be done in terms of preventive action. The rubber wedge could have extra rubber, the cut steel edges could be folded under the belt, the edges of the belt could be protected by a circumferential belt cover, and the belt-to-belt adhesion rubber could be strengthened and made more durable (rubber chemistry and materials processing). The more specific risk frequency data could be obtained from historical and ongoing adjustment data, data trends, and adjustment classifications. Warranty data is also a good risk indicator. The danger became manifest in terms of collision and injury claims. The danger decision was made and a voluntary recall program was initiated. A government-mandated recall program could have been made after a risk evaluation and a determination that an unreasonable risk was manifest.

(3) Hazard prevention

As the foregoing tire failure analysis indicates, the hazard identification should be an interactive process, successively probing ever deeper into causation. This procedure, by itself, might uncover simple, low-cost, and effective remedies. The principal objective is to locate alternative designs to eliminate or remove hazards. This is a hazard prevention objective that can be very effective during the early stages of design, development, and testing. If the changes can be made quickly and easily, there may not be a need for a detailed risk assessment and decision-making as to danger. In this respect, it is an exception to the rule of the basic trilogy (see Table 3.1).

After significant meaningful experience in consciously applying the basic trilogy method or its hazard prevention exception, the safety analyst may develop something of a blind-sight capability (no overt focus, seemingly instinctive, or neurally primitive) for detecting early symptoms of error or failure and assessing probable cause. For example, a quick tire inspection may reveal belt edge hazards by feel (touching the tread edge grooves for irregularities), vision (small holes in the shoulder of a tire resulting from

Table 3.1 Hazard prevention (top down)

Inquiry level	Hazard type	Tire example	Collision example
Top tier (what happened)	Apparent hazard (early classification)	Tire failure (belt separation)	Crashworthiness (crash pulse uneven)
Intermediate tier (possible failure mechanisms)	Probable hazard (causation search)	Crack propagation (loss of adhesion)	Front-end structure (weakness where strength needed)
Base tier (why it occurred)	Operative hazard (remedy implied)	Crack initiation (belt edge cutting rubber)	Cross-member strength and configuration (crash energy distribution)

belt edge fraying and cutting), and sound (tapping a truck tire for the sound of an air pocket bulge of a tread separation). Determining the in-depth causation of these problems may seem to be a rather skillful exercise. This is quite different from what a tire changer, inspector, or adjuster is told to look for and how to classify tire problems. The difference is in what the causation can signal in terms of pinpointed effective correction of a safety problem.

In essence, there is insufficient in-depth information on the hazards that are identified in most accident reports, warranty claims, adjustment data, repair orders, parts replacement invoices, logistic or field reports, and customer complaints. The data is insufficient because it is too general for precise corrective action. A hazard prevention inquiry or probe is often necessary to identify specifically (pinpoint) the operative hazard that clearly suggests what must be corrected. This enables changes that are least disruptive, most financially and technically feasible, and more effective than broad-range efforts to hit an ambiguous target.

(b) Decision models

After the amount of risk has been estimated, some decision has to be made as to its acceptability. What criterion (test) should be used to determine whether there is an unsafe condition, a process danger, or a product defect? In essence, this is the ultimate decision as to whether there is a danger that requires warning, safeguarding, redesign, marketing and use limitations, or other actions. A management selection must be made as to which of the commonly used risk decision models should be used. The following list of decision models is presented in the order of their strictness, the higher safety requirements down to the least constraining.

(1) The absolute control model

This is a zero tolerance of permissible risk based on the social utility theory. The risk may be acceptable if it serves a useful (utilitarian) function that is otherwise not available. But something that is inherently dangerous is always suspect. For example, mountain lions cannot be tamed, serve no useful purpose in a residential neighborhood, and are a high risk when not in an enclosed cage. A military weapon may be somewhat inaccurate, have a high risk or threat potential, have a different function from any other weapon, but still serve a social purpose. The same claims may be made for an industrial or commercial product or process. There may be some highly specialized automotive vehicles that, arguably, could pose high risk but have great social utility. In essence, this is the 'no other means' argument.

(2) The excessive preventable risk model

This is based on the theory that significant threats to humans, property, and the environment should be avoided if at all possible. The life-saving potential is considered a paramount consideration. This model is presented in the question: 'Do you agree that no risk of serious injury or death is acceptable if there are reasonable means to mitigate the risks or prevent the danger?' If the risk is preventable, and the risk is excessive, it constitutes an impermissible danger. In essence, the question is what is excessive and what is preventable.

(3) The expectations model

This is based on customer satisfaction theories, i.e. the customer is king, if we are to survive! The expectations may be of another company (vehicle assembler), the user (driver or consumer), or entity (government agency). Unfortunately, such expectations may be implied as well as express, are often quite subjective and ambiguous, and may be discovered after the fact. The first awareness of such a design requirement might be the customer saying: 'We did not expect this component to fail under the peak loads experienced by our product, we expected it to work properly in our product.' The consumer expectations model (test) may rely upon a consumer who has no basis for expectations of a product involved in a confounding and complex environment. However, the expectations model is popular because it empowers the purchaser to express dissatisfaction and to have redress before there are long-term consequences in the form of an adverse reputation or choice of alternative supplier.

(4) The compliance model

One of the most common models is compliance with all laws, regulations, contract requirements, design specifications, trade standards, technical recommendations, provisions in company design manuals, and other guidance documents. This is a comparatively simple process, often with clear, specific, and defined objectives. This model is highly favored by those who dislike ambiguity. However, standards cannot cover each and every possible hazard, may be deceptive, and are often only trade consensus documents presenting some easily achieved minimum. They do serve as a good first step in recognizing problem areas. Standards committee members could help to detail what has prompted their attention and work. In most situations, standards should be considered mandatory and usually a helpful review can be made of related professional association standards. Standards provide an additional decision-model utilized for an assurance that decisions are prudent and proper under the circumstances. In essence, compliance may be only one model of several that should be utilized.

(5) The comparative model

This model is used to compare one product or process with another. Is it safer than a competitor's product? How does it rank among a group of useful products or estimates of morbidity and mortality from naturally occurring events? It may be a deceptive model if the product comparisons and data are carefully selected or weeded-out and the estimates are speculative or inaccurate. It provides little guidance during advance design when realistic data are not yet available to use for a comparison. This model is generally disfavored based on past experience.

(6) The informed assumption model

This model presumes, under a safe-use doctrine, that an individual or entity will voluntarily engage in risk-taking behavior, in an informed manner, and with due care in terms of safety. Some factors suggesting voluntary assumption of a known risk are: *incentives* such as hazardous duty pay for performing tasks with special risks, *obviousness* of the danger sufficient for a reasonable person to comprehend that substantial risk is being undertaken, and *inducement* by virtue of special instructions and retention of employment. An obvious danger might be speeding on a curvy mountain road, performance of a task requiring the use of personal protective equipment, or continued use of a product after notice of a safety defect. It is arguable whether the assumption of a risk is voluntary after threatened job loss. Questions also arise as to whether the risk was fully recognized and appreciated. However, it can also be argued that people accept all kinds of risks every day and everywhere. In essence, most people voluntarily assume a broad range of risks and it is their responsibility to learn and apply safe-use principles. Under some interpretations, this could be a very lax criterion as to danger and could tolerate many otherwise unacceptable risks.

(7) The shift-the-burden model

This model is usually a defensive and adversarial technique for shifting responsibility to others. The design engineer may be asked to 'prove it' and 'convince us' at a stage where he has insufficient data to justify the risk as a danger. If the burden of proof is shifted to others, there is rarely an admission that a danger could exist. If the proof is never sufficient or clearly presented, could a true danger go undetected and uncorrected? In essence, shifting the burden may be shifting the remedy to a later stage in design and development where changes are less favored and more costly.

(8) The developmental risk model

In some countries, it is believed that there should be a special exception for products still in the early developmental stages where hazards have not been discovered and risks have yet to be assessed. This is based on theories that new developments should be encouraged and early mistakes should not be punished. At worst, anything goes. At best, the exemptions are conditional and limited. For design decision-making, it offers little in the way of guidance. In essence, some other decision model should be used to determine danger and the need for preventive action.

(9) Design models

These decision models are tentative working hypotheses for preliminary design analysis purposes. They may categorize hazards as having low risk, minor risk, moderate risk, significant risk, or high risk. The purpose of the qualitative assessment is to determine worst-first (pareto principle) allocations of design improvement effect. Variations of hazard identification may be the failure mode analysis or the preparation of diagrams showing interactions between groups, functions, activities, or components. In some companies, quantitative estimates of each may be made on a scale of 1 to 10 or in terms of failure per million operations or cycles. These are often quick judgment estimates based on limited, uneven, or missing data. They do serve to focus attention on possible hazards, risk reduction by alternative designs, and serve a valuable system engineering assurance function. The value of such an analysis depends on the engineers and the support they are given within the company's organizational structure. In essence, the design model provides for rather speculative danger-decisions rather early in the design process in order to focus attention where it will do the most good in terms of risk reduction. It is generally not a pass–fail criterion but a classification of levels of risk.

(10) Mental model of risk

There are often differences in opinions as to perceptions of risk. These differences are reflected in the mental models that determine the coding or reference points used for subjective judgment. The judgments can be justified by rationalization and the use of selective evidence, but are usually the result of self-interest biases (company-compatible or from some set of personal values). They often become ingrained and unchangeable by positive reinforcement. Thus, there are multiple risk assessment techniques, methods, and outcomes. In essence, there may be an individual or personalized mental model of risk that overcomes or clouds the use of other commonly used decision-making models. The use of personal subjective judgment instead of commonly recognized and utilized decision-models can result and has resulted in situations having serious consequences to companies and to society as a whole.

(c) Balancing risks

There are many competing considerations, other than just design feasibility, that are used to evaluate the true risk of products, processes, and services. These considerations may give greater weight to political, social, financial, or special interests. There may be broader definitions of economic (cost) impact that include community interests. There may be practical (technical) considerations that include employment (jobs) in a domestic market. Thus, special balancing or weighting may be given to considerations deemed important.

A fairly common balancing procedure is the risk–benefit test; that is, do the benefits outweigh the risks? Another is the risk–utility test; that is, is the utility of the product or process greater than the gravity of the danger? The objective is generally to overcome opinions that there is unacceptable risk.

There are several concepts that have been used to affect the balancing of risks. For example, the 'managed risk' concept assumes that controls will be exerted to reduce the risk. Who will manage to assure safe use and what is the probable outcome? There may be 'unavoidable risks', but there will be contractual informed consent to accept the risk. If 'reasonably practical' measures are undertaken, there may be hopes that this will control risk exposure. The existing background risk serves as a reference point for determining any excess risk in the 'relative risk' concepts. In essence, there may be many novel concepts and considerations used to lower the overall risk estimates and declare that what has been called unsafe is really safe. This differs from alternative design remedies in terms of using novel concepts and balancing considerations to dress, convert, or optimistically predict that substantial risk may be acceptable. The concepts may be meritorious in other settings or another context, but may hinder the use of alternative designs or other preventive remedies. In essence, caution should be exerted when dealing with subsequent or supplemental risk evaluation procedures.

(d) Combining risks

Risk estimates are generally formulated for one hazard in one product or process. But there may be other hazards and risks in the product, so a combined or total risk estimation may be desirable.

Block, link, event tree, or logic diagrams may be used to determine the relative independence of a hazard or its linkage to other events, hazards, and risks. This alone has been very useful in some overall risk assessments.

Redundancy refers to having more than one means (elements) in a system to accomplish a given function. Assuming that identical (hazard) elements are in parallel and both are active, the risk would be doubled. If one element is on standby status (inoperative), the risk is not doubled. Thus, a system design perspective is important in total risk estimation and control.

The risk may be present or absent if the power is on or off, if the system is operating or not, or only during maintenance, service, repair, or disposal. Thus, the total risk should be calculated for the time the risk is actuated (alive) and for the life cycle of the product or system.

Similarly, there may be two different hazards and risks in one product or system. The question might be whether the risks are additive, multiplicative, or otherwise modified. What is the operational relationship between the risks?

The safety analyst should remain aware of possible combined risk problems and consult a reliability engineer as to the total or overall risk implications.

For further information, see Amstadter (1971), Von Alven (1964), and the Chemical Manufacturers Association (1985).

Caveat: Estimating risk is not an end in itself, it is merely a means to allocate resources rationally to reduce risks to acceptable levels.

(e) Biological risk assessments

Sophisticated risk assessments are performed on biological systems, with various national governments providing specific guidance on the methodology, terminology, criteria, and specific objectives to be used in the analysis. A risk assessment may be made based on animal studies of a chemical that could cause harm to humans. Regulatory action could result from the findings, so there is often intense criticism of such studies. They may be considered legally insufficient, lacking in peer review and citizen input, or based on unacceptable uncertainties. They may serve to promote or inhibit governmental action, foster a change in permissible exposure standards, or encourage chemical formulators (manufacturers) to alter their product stewardship or research programs.

The result of risk assessment is somewhat akin to a public health summary statement that at one time there were approximately 50,000 deaths in automobile accidents, out of a population of about 225 million persons, yielding a yearly risk of 22.2×10^{-4} or a chance of about one in 4500 of dying in an automobile accident. Of course, this assumes that the death rate and population will remain constant in the future, whereas the population is increasing, and hopefully design safety improvements are reducing the number of deaths.

Of particular interest is the following definition (National Research Council, 1983):

> Hazard identification is the process of gathering, organizing, and evaluating all data that may reveal the types of adverse biological effects that may be produced by the substance under evaluation and the conditions under which they are produced. The evaluation should be critical, and include a characterization of the strength of evidence that the substance causes specific effects, the underlying biological mechanisms that contribute to those effects, and the extent to which observations in various biological models (e.g., animals, microorganisms, etc.) can be said to apply to humans. . . . All uncertainties need to be specified.

Unfortunately most risk assessments are not that sophisticated.

The risk assessment stages are usually hazard identification, dose–response assessment, exposure assessment, and risk. In essence, the hazard identification is determining whether an agent causes a health condition, with causation a major issue; the dose–response assessment is determining the effect of increasing levels of exposure to an agent; the exposure assessment is estimating intensity, frequency, and duration of exposure; and the risk categorization is estimating the incidence of a health effect.

The findings may be categorized as a *de minimis* (insignificant) *risk*, an *acceptable risk* (less than one-in-a-million chance), or in terms of *relative risk* (the excess over the background risk). If the risk is categorized as *serious* or capable of widespread harm, it may warrant immediate high-priority action.

Thus, risk assessments vary as to their purpose, sponsors, and type of information being processed. But there is increasing sophistication in methodology, data accumulation, and effectiveness whether in engineering design, human health, or for environmental concerns.

Caveat: In terms of world trade, there are important differences in determining what constitutes a danger. The European Commission has adopted the 'precautionary principle' in risk assessments, so a danger may exist whenever there are reasonable grounds for concern over potentially dangerous effects on humans, the environment, animals, and plants. In the United States, the Supreme Court has decided that there must be reliable scientific evidence to prove a harm. Thus, the United States requires what has proved to be a high and difficult level of proof. Whereas, in the European Union, a conclusion that a danger exists can occur despite insufficient, inconclusive, or uncertain scientific evidence. It is anticipated that there will be, eventually, some harmonization of such risk assessment criteria, but cautions should be exercised in this and other safety-related criteria.

In regard to the document *Reducing Risks, Protecting People* (Health & Safety Executive, 2001): 'the revised version reaffirms the importance of quantitative and qualitative risk assessment in our decision making process' (personal communication to the authors of this book from David Rickwood, Risk Policy Unit, HSE).

4 Human error control

(a) The blame game

Everyone is familiar with statements that some automobile driver acted in a careless, inattentive, and reckless manner. Such general opinions clearly suggest human fault and error. They serve to blame the driver and tend to close any further inquiry as to causation either of the behavior or of a possible accident.

Unfortunately, such generalizations may not help in defining appropriate and effective countermeasures. Something more is needed for adequate human error control or accident prevention.

The following examples illustrate the blame game and the superficial remedies that could result: excessive speed for road conditions, failure to decrease speed in a construction zone, following too closely, failure to comply with traffic signs, failure to stop at a red light, failure to yield to a pedestrian, failure to properly secure a load in a truck, failure to yield the right-of-way, and driver made a left turn before the intersection cleared.

Slightly more detail on causation might change the blame for a driver who lost control of a vehicle, which then rolled over. The extra detail is that the event was preceded by an inadvertent airbag deployment. The deeper detail is why the airbag deployed and how the airbag system could be remedied. Similarly, consider the careless individual who suffered a wrist injury when a jack failed and a van fell on his wrist. He was considered careless for not blocking the vehicle, using the jack incorrectly, and placing his wrist in a position where it could be injured. Another perspective is that the jack collapsed because it did not have a mechanical safety or stop pin to prevent it from dropping or releasing the load when the level or pressure of the hydraulic fluid was too low to support the vehicle. The caveat is simply that blame should not be assessed on quick conclusions or superficial opinions that do not suggest meaningful preventive remedies. Such a bias may preclude a more detailed and thorough analysis that could be more useful, fair, and meaningful.

Thus, the approach to human error control must be commensurate with the complexity of human behavior, whether it is in design, manufacture, use, or repair of automotive vehicles. Some meaningful depth of analysis, use of available techniques and knowledge, and effective countermeasures are desirable. Simply saying that an act is incorrect, wrong, and a foolish move does not demonstrate any insight as to causation (the why) and possible remedy (what to do about it). The blame game alone may only serve to allow human error to persist and accidents to continue.

(b) Basic approaches for human error analysis

There have been some estimates that human error is involved in 50% to 90% of all accidents. This may be due to the simplicity in just blaming the person involved in an accident, the limited forced choices available in some accident checklists or police report forms that are completed almost immediately at the scene of the accident, or the accident investigator's basic philosophical approach to human error. People do make mistakes and this seems to be an inevitable certainty. But it constitutes a serious challenge as we strive to achieve ever higher requirements for complex products, processes, and services. The ultimate question is: how can we significantly reduce and control seemingly intractable human error, mistakes, and human failure?

The selection of a philosophical approach usually determines the outcome in human error evaluation, investigation, reconstruction, cause analysis, and fault determination. Philosophy and value systems produce a bias and focus that can inhibit an analysis or make it more productive. The five most common approaches are as follows.

(1) *The passive approach*

There is a philosophical perspective that fatalistically considers human error as inevitable in an imperfect world populated by imperfect people. In other words, there is a belief that errors will happen and there is little that can be done about it. This can breed a do-nothing, indifferent, or passive reaction to human error. For example, within a company there may be a surveillance of monthly summaries of product returns, warranty claims, injury complaints, product discrepancies, or component defects. The analyst with an essentially passive approach may merely classify the problems into categories having limited or no corrective action possibilities. There may be no consideration or personal interest in further inquiry or possible action as long as the data remain within some subjectively determined acceptance limit or band. It may be assumed that any action is someone else's responsibility. The risks are tolerated and may be low for short shelf-life or short duration service life products. But, the risks for long-life products may be far greater. Sudden excursions of data may be neglected even though they have unacceptable quality, cost, or market implications. In essence, the product surveillance system may be intended to be an early warning system for in-field or customer problems and to help initiate prompt improvement, but a bad attitude can frustrate the process.

From a business standpoint, a passive approach to possible problems is a high-risk situation. This is particularly true where there are already differing opinions on the corrective actions that are technically and economically feasible. The symptoms of a passive attitude are rationalizations or excuses that include words such as 'reasonably expected errors', 'reasonably safe under the conditions', 'harmless errors', 'inherent errors', and 'normal accidents'.

There are also ethical problems with a passive approach. Various cannons of ethics state that 'the engineer will have proper regard for the safety, health, and welfare of the public in the performance of his professional duties' (NSPE). The professional's paramount objective is to protect the health and safety of people (ECPD). The question is whether passive indifference is an ethical violation or whether failure to inquire, take the initiative, or investigate is a moral issue in terms of basic value systems required in a civilized society. If so, a different philosophical approach to human error may be needed.

(2) *The behavioral approach*

Another approach to human error is to focus on undesirable behavior and unsafe acts, then attempt to develop safer people, achieve zero defects, and strive for error-free human performance. An effort could be made to motivate workers, to encourage them to act responsibly, and to help them develop safer attitudes. This may be accompanied by special training to improve job skills and hazard recognition, providing more detailed instructions and procedures, and encouraging closer workplace supervision. It may include an assessment of the psychological and physiological functions required for better job performance, job assignment, and personnel selection. In essence, improved behavior can reduce human error to some degree.

Short-term improvements can be expected from a purely behavioral emphasis. But it has been found that the cost can be substantial for continual efforts to maintain individual motivation, closer supervision, adequate training, and to perform related job performance evaluations. It may be very difficult to alter some individuals' perceptions of risk and risk-taking behavior. The essential flaw is that the residual human error may remain too high, unacceptable excursions of human error may occur unexpectedly, and some human errors seem intractable and resist reasonable attempts at improvement.

In terms of accident reconstruction, the cause analysis might be very simple because the behavioral fault would generally fall on the proximate individual whose behavior could be described in a blameworthy fashion. The accident victim who is injured may be unable to speak or defend against allegations of unsafe behavior. No-fault workers' compensation and other insurance schemes generally militate against findings that blame persons or entities other than those injured. In essence, this approach is one-sided, since there may be other causes and cures for human error that should not be neglected.

(3) *The situational approach*

Instead of inaction (the passive approach) or blaming the victim (the behavioral approach), there could be an attempt to blame the situation in which the accident occurred. The fault might be perceived as in the work environment, the unique overall circumstances of the accident site, possible group interactions, improper workplace management, or the tools provided for the tasks required. It could include the failure of a product to meet the user requirement under the circumstances of an accident. This is a fairly broad perspective as to injury and error causation, but the focus is on the particular situation, accident site, or environment. It may be so broad as to include concepts of human rights and social justice. It may be so broad that practical and specific remedies are not technically, economically, financially, or politically feasible.

(4) *The product design approach*

The design engineering focus has gradually broadened to include more user-friendly designs, driver-assist devices, and consideration of human factors criteria. It now includes some conscious effort to design for ease of manufacture and compatibility with operations in the product life-cycle. This broadened design responsibility serves to place some personal responsibility on individual design engineers for acts of commission or omission in regard to predictable human error. In essence, it shifts everything to the design engineer.

This approach is still dependent on available specialized knowledge, relevant data, appropriate design criteria, productive test methods, and access to aftermarket problems. Understanding field use and the complexity of human behavior are often impediments to this low-cost approach to human error control.

(5) *The multifactoral approach*

The most sophisticated approach is that of multiple causation. There is usually more than one cause of an accident or a human error. Any one of the causes might provide an opportunity for preventive action. Prevention of only one cause assumes that it is substantial or necessary in the chain of events.

It avoids the tendency to make quick and simple snap opinions as to causation. It reduces the bias from preconceived opinions that act as gatekeepers to tailor selectively or accept only facts that fit the initial hypothesis. In other words, frequently there is a tendency to believe that a person is guilty and then gather the facts to prove it. However, full data collection is generally necessary for effective human error control and injury reduction.

(c) Illustrative errors

(1) *Judgmental errors*

General examples

The vehicle driver may have difficulty judging his speed and that of other drivers without looking at a speedometer. Fast-moving traffic may increase what is perceived to be a safe speed and the driver may then underestimate his own speed. In other words, human judgments are influenced by the surroundings or environment of use.

Underinflated tires may be considered a human error, a failure to maintain the tire in the recommended condition. Some factory adjusters claim that the driver should check the tires every day before operating the vehicle. This is something that is not a customary practice because the consumer has been the subject of advertising claims, knowledge that tire design and manufacture have improved rather dramatically, and that tire inflation and measurement equipment is no longer available or is relegated to a remote location at gasoline (fuel or petrol) stations. People rarely check the tire pressure regularly.

Low-profile tires make it difficult to perceive low inflation pressure conditions. Various studies report that 25% to 30% of all vehicles have tires that are underinflated by about 8 psi from the manufacturer's recommendations. An underinflated tire may overheat in the shoulder of the tire and overstress the carcass sidewall, leading to premature failure.

A solution to this predictable human oversight or judgment error is the tire pressure sensor, microprocessor, and instrument panel warning of low tire pressure. The system may be *direct* by having in-tire battery-powered radio signals transmitted to the 'cockpit' receiver. A lower cost *indirect* system utilizes the antilock brake system to count wheel revolutions. A low pressure tire spins more quickly than a fully inflated tire, the tire rotations are compared and significant differences serve to actuate a warning light or display readout in the instrument panel of the vehicle. This compensates for errors in judging appropriate tire inflation pressure.

Proximal referent

Human judgments are often made in relation to some reference or anchor point that may be proximate, temporally or spatially. For example, an automobile driver tends to adjust his vehicle's speed to that of the adjacent vehicles in his lane and adjacent lanes. Rather than look at the speedometer to determine the actual speed, it is easier just to look around and follow the others. It is the use of a proximal referent.

There may be an implicit assumption or rationalization that the other drivers have more information and familiarity with the roadway and traffic conditions. Perhaps it is a passive judgment rather than an active independent judgment: one that is considered good enough for the situation and the immediate needs of the driver. Any errors are group errors.

Judging vehicle speed from non-moving objects along the roadside is a more complex process prone to even greater human error. The decision-making is based on more *remote referents* and requires extrapolation and abstract reasoning, particularly if there is a lack of familiarity with the roadway and its surroundings.

Even more remote, temporally or spatially, is the use of past experience in judging vehicle speed. The referent is expectations based on *past memory*, sometimes involving the interaction of several brain areas. The speed may seem right, but it can be very incorrect.

Judging vehicle speed is always a *comparative process*. Looking at a speedometer is immediately followed by comparisons involving active memory imagery; looking at other moving vehicles involves comparing the vehicle's speed with that of others, watching the trees go by and comparing it with expectations as to speed, or remembering past experiences and comparing present speed indicators.

Judgments as to vehicle speeds and resultant human error illustrate how such errors could be classified. Some are *correctable without damage*, if the vehicle could be decelerated and travel resumed at a legal or more reasonable speed. Perhaps this is an example of good judgment. It may be *correctable with damage* if a speeding (police) ticket or collision results before the speed is reduced. This might be considered a bad judgment. Stepping on a brake pedal too late to avoid a collision might be an *uncorrectable human error* if timely evasive steering action could not have been undertaken. This would be a case of very bad judgment.

To improve human judgment, a more proximate indication of vehicle speed could be made by a heads-up display (HUD) projected on to the windshield on demand or during overspeed conditions. An audible warning device might serve to alert the driver of excessive speed, which might be over the legal speed limit, the customary speed limit, or a speed limit commensurate with the driving conditions. Another example is cruise control mechanisms designed to keep a constant vehicle speed. The driver sets the speed, the vehicle keeps it within a small range, and the driver can remove his foot from the accelerator pedal and relax. Some relax too much, even on a relatively straight, uncluttered rural road. This convenience device is not well adapted to urban traffic that requires almost constant changes in vehicle speed. Human judgment in the driving task is almost always needed and higher vigilance is generally higher safety.

A subjective judgment problem existed for self-propelled elevating work platforms, such as scissors lifts. They are of narrow width and could tip over at extended heights. Workers were cautioned to use them only on hard and level surfaces, not on uneven or sloping ground. But an asphalt paved surface near buildings must have rain (water) drainage, thus a slope or inclined surface. The self-propelled platform might be driven in a partially extended (up) position and hit a pothole with one of its small-diameter wheels. Such vehicles are used on slanted grass surfaces and uneven terrain. There are special rough terrain lifts that look somewhat the same. The question then becomes: how much of a slant is too much for a particular elevating work platform? Some surfaces look (perceptually) flat to the operator, but are not. Thus, there were a number of tipover or turnover accidents involving serious injury.

To reduce the risk of tipover, lifts were equipped with manually or hydraulically actuated *stabilizers* or side legs with pads that could contact the ground surface, thus providing greater width for the vehicle base. A wider wheel base could do the same job, but the articulated stabilizers preserved the narrow width necessary to move the lift within buildings. Later, powered *outriggers* were used both to support and move the lift into a more upright position when on sloping or uneven ground surfaces. Then, lifts were equipped with tilt monitors, upset warning alarms, or slope sensors (such as gravity-actuated pendulum switches, rather than bubble levels or differences in outrigger leg hydraulic pressures). They alerted the operator if the lift approached the topple point and some also either automatically lowered the platform or deactivated all powered functions. Pothole protection was afforded by long rails on each side of the lift that lowered to contact the ground surface and raised (retracted) as the work platform ascended.

Similarly, a boom-type aerial bucket mounted on a truck chassis may have wide-span hydraulically actuated outriggers which raise the tires off the ground. These stabilize the bucket or crane. Audio and visual warnings may sound in the cab when the limits are approached in terms of boom length, angle, working radius, and actual load. All such devices compensate for human judgment errors in safety-critical situations.

It is of interest that adjustable outriggers were used on self-propelled cranes more than 80 years ago. They greatly extended the side lift capacity (tilt axis) of locomotive and truck-mounted cranes and eliminated the 'give' of spring suspensions and pneumatic tires. Many outriggers are still in the form of aftermarket kits or optional safety devices. As a general rule, safety devices are slow to be adopted because the easiest solution seems to be better safe work practices and procedures.

The operator of a large straddle crane may have his vision obstructed by the load, his perception impaired by the distance to all four legs of the crane, and may manifest a judgment error as to travel path speed relative to moving obstacles. Workers were crushed under the wheels of such cranes. The accident cause was generally considered to be a judgment error by the operator and lack of attention by the worker. The remedies consisted of wheel guards (bumpers), audible travel alarms (warnings), two-way radio communication (operator and ground crew), and panic (pressure or proximity) bars on the wheel guards to stop the crane movement. Poor judgment is foreseeable, but the consequences are preventable.

(2) *Anthropometrically induced errors*

A driver attempts to push one button on the instrument panel, hits another, and is surprised and chagrined at the error of the device. This is called an aiming error or a 'fat finger' overlap keyboard error. The fat finger may be anthropometrically large or just clumsy in use, without direct visual–motor brain guidance.

Small females may *crowd* into the steering wheel (airbag) attempting to get a better view and reach the pedals with their legs. Average-size drivers may have *cramped* legs, after long drives, because of pedal depression forces, pedal positioning, and the driver leg position.

Males with tall upper bodies may bump their heads against the roof, those with long legs may awkwardly depress the accelerator pedal or brake pedal, and those who are pleasingly plump may push backward and tilt the seatback to gain living room and comfort. The key safety problem is maintaining a correct eye height and location, so that driver visibility is not compromised.

To accommodate drivers of different sizes, the seats became more and more adjustable (manually, powered, and with position memory). The accelerator pedal and brake pedal became adjustable. The steering wheel became adjustable (telescoping up and down and tilting). There have been attempts to design a moveable instrument panel and steering wheel, in order to better position the driver. The fixed location seat, keyed to eye position, might serve to locate the driver properly, then the instrument panel, steering wheel, and pedals would move toward the driver as part of an occupant restraint system. However, there are issues relating to individual comfort positions, open compared to closed occupant spaces, the use of automatic positioning by eye height sensors, and customer acceptance. In essence, accommodating differences in human sizing (anthropometrics) is a difficult design problem, particularly for small economical vehicles. But each misfit creates a situation that increases secondary induced human error.

Another example of anthropometrically induced error is in the deployment of airbags. Size does matter, since there is a critical distance relationship between the driver's or passenger's head location and the location from which the airbag is inflated. Small children are easily mispositioned and this can result in serious injury.

Seat sensors may determine the size or weight of the occupant, whether the occupant is out of position (OOP) on the seat, and the eye position. The airbags may be triggered or not, have staged inflation or variable rate power, or otherwise adjust to the size and position of the occupant. Improper inflation pressure and timing might be machine error, improper occupant positioning may be human error, and unusual occupant sizing may be a system design problem searching for a good design compromise. Errors abound when anthropometrics are not fully accommodated because there are no driver license restrictions for the anthropometric extremes. A cramped leg driver, a driver who becomes easily fatigued because of misfit problems, or a driver who has an eye location that restricts the field-of-view to the side or rear is a good candidate for human error and accidents.

In defining possible anthropometrically induced problems, the error may be considered primary or direct, such as a long-legged driver whose foot becomes mispositioned on the brake pedal and slips off during a panic stop. It may be a secondary or indirect error, such as when misfit problems place special demands on the driver which fosters perceptual narrowing of the visual field and driver fatigue, and this induces secondary driver errors.

Anthropometric problems are more likely to become manifest on industrial, construction, recreational, military, and special-purpose vehicles. In regard to a recreational vehicle: 'Our complaint is about the convertible sofa that is actually 10 inches shorter than the one shown in a picture in (the) brochure and too short for an adult to sleep on' ('Buyer's Remorse', *Trailer Life* (magazine), December 2001, p. 16).

Where such problems exist, it may become a matter of selecting a driver that fits the equipment. Although this may be difficult, as can be understood by anyone who has seen a child driving a farm tractor, there are going to be poor choices and then attempts to modify the equipment. The rationalizations are: it's a matter of consumer beware, look before you buy, and make your needs known to the retailer or supplier. Bad fit is creating the potential for misuse, accidents, and damage.

Not all vehicles are equipped with telescoping steering wheel columns, tilt steering wheels, adjustable seats, adjustable mirrors, and adjustable foot pedals to accommodate human dimensional variations.

(3) Oversteering errors

Steering problems are a concern if the driver would like the vehicle to move in the direction desired, in a fairly precise manner, and be responsive to the steering wheel input. The driver might be concerned if the vehicle tends to wander within the traffic lane (a loose steering). The vehicle may be resistant to steering on a curve (hard to steer), which is called understeering. Worse yet is for the steering to be too sensitive and the vehicle too quick to steer. This is called oversteering and could result in a spin-out. These conditions are often considered poor steering by the driver or human error.

There are many factors to be considered in terms of desirable steering responses and the more expensive vehicles tend to have the features, devices, and attention to detail that results in laudable vehicle handling or 'good manners' on the road.

One European test is to evade a deer (by simulation) on the road by a turn in one direction and then a turn in the other direction (an S-turn maneuver), at a given speed, and without a spin-out, rollover, or loss of vehicle control. This is similar to a common accident scenario in which a vehicle drifts off the road, and the driver attempts to steer back quickly on to the paved road surface. The driver may overcorrect by too much steering input, in which case the vehicle enters the road at too great an angle and moves diagonally across the roadway. The driver then turns in the other direction to head the car down the road and does this quickly in an attempt to regain control over the vehicle. The vehicle undergoes an S-maneuver and tends to sway or roll on its longitudinal axis. If a tire deflects sideways and the rim digs into the pavement, it could trigger a rollover situation. The driver may be blamed for the following human errors: drifting off the road, overcorrecting twice, and failure to control the vehicle properly.

Some automobile experts may fault a vehicle's soft suspension as well as the oversteering error. One common remedy was the torsion bar or rod that attached in its middle to the frame, with each end attached to the wheel support or suspension members. Sway forces on one side (wheel) were transferred to the other side (wheel) to better position the wheels and balance the loads. In hundreds of patent variations, these were known as stabilizer bars, anti-sway bars, and anti-roll bars. In addition to the front wheels, torsion bars were attached to the right rear wheel supports to prevent wheel hop and to both wheels for added vehicle stability.

In a cornering or turning maneuver, there should be a compensating weight transfer from the side (of the turn) to the outside wheels. The stiffer the springs and the roll bar, the greater the weight transfer to the outside of the turn (that is, less roll, rollover, and oversteering).

More recently, electronic stability systems, using yaw sensors, monitor the vehicle direction and selectively apply braking action to return the vehicle to a stable direction The device compensates for oversteering, whether driver- or vehicle-induced, so the driver will not lose control and the vehicle spin-out. In essence, oversteering errors have been a critical safety problem, but the use of anti-roll torsion bars and electronic stability systems has compensated and limited the consequences of the error.

(4) Braking errors

A passenger car driver applied the brakes of the vehicle, but it skidded out of control on a sheet of ice. The skid could have been on foreign material, asphalt bleeding, or wet pavement. In any such case, a police investigator could quickly conclude that the driver was proceeding at an unsafe speed for those conditions; that is, it was simply driver error.

Different vehicles have different stopping distances and those distances are affected by the road slope and road surface roughness. Too great a stopping distance might result in a front collision; too short a distance might result in a rear collision, since braking often occurs on a crowded roadway or in stop-and-go traffic jams. In other words, the driver must exercise caution in a rather complex interaction of objects moving in different directions at different velocities on different surfaces, then making anticipatory decisions on steering, acceleration, and braking.

Additional predictive information might help the driver's decision-making process, so the braking might not be a last-minute decision. This is the reasoning behind the development of collision-avoidance devices, which search, sense, compute, and display warning information. One such device is a simple instrument panel display that, when lighted, says 'warning collision'. Other collision-avoidance devices provide a warning, but if the warning is not heeded the vehicle will automatically brake and, if the adjoining lane is clear, it will automatically steer to avoid the object in front of the vehicle. Unfortunately, the prediction problem is difficult with inaccuracies and reliability problems, so complete reliance on such systems may be unjustified at present. However, with appropriate human factors research, software development, faster response rates, and further preproduction road and fleet testing, the system could be of material assistance to the driver.

It is of considerable research interest that the human brain has a primary focus on anticipation (prediction) in terms of information pattern (system) completion. This may be why braking error rates are surprisingly low for such a common task. The brain perceives a pattern of activity on the road and predicts important results, thus alerting the driver in time to apply the brakes properly.

Loss of vehicle control on wet roads has been reduced by antilock brakes, so the driver need not pump the brakes in a very rapid fashion. Similarly, loss of vehicle control is reduced by dynamic or electronic stability systems that permit brake application to continue as the vehicle begins to spin or slide out of control. Proper grooving of tire treads reduces hydroplaning and loss of control on very wet road surfaces. Thus, there are devices to reduce human error during the braking of automobiles, but further research and development of human-aid devices could substantially reduce braking errors.

During normal braking, the driver is constantly assessing the situation and adjusting the pressure being exerted on the brake pedal to modify the slowing, retarding, or stopping action. For mechanical linkage brakes, vehicle deceleration is roughly directly proportional to the force exerted on the brake pedal. For power-assisted brakes, deceleration may be proportional to the pedal travel or distance depressed. In an emergency situation, considerable force can be exerted on the brake pedal (more than twice the human body weight), and in some vehicles, the force applied has been sufficient to bend the brake pedal arm. The brake pedal travel may be in the range of 2 to 4 inches (5.1 to 10.2 cm) and should not require more than 90 lbs force (the 5-percentile female limit). The deceleration may exceed 1 *g*. Fortunately, drivers generally adjust quickly to differences in pedal depression and vehicle deceleration, but they may make errors in assessing the effects of road surface, ice and snow, underinflated tires, and the load being carried.

Incorrect braking of tractor trailer rigs can result in jackknifing (tractor rear wheels locked and cab rotation to the trailer in one or two seconds), trailer swing (trailer wheels locked and a slow change of angle between tractor and trailer), and load shifting. If the loads are secured and head boards are provided, the truck should not have trouble with decelerations of about 0.5 *g*. Some trucks tend to tilt on their springs during braking and cornering. If there is a combination of cornering, braking, and load shifting, the truck or trailer may tip over sideways. A truck may roll at 0.4 *g*, whereas a passenger vehicle may require in excess of 1 *g* to roll.

There may be weight transfer (more tire-retarding force on the front than the suggested weight distribution on the axles), resulting in a 'nose down'. This may be a problem for trucks if the weight is shifted to non-braking wheels, but for motorcycles and motor scooters it is far more serious since the rear wheels ordinarily provide only about a third of the necessary deceleration. Thus, if the front wheel locks, the motorcycle may go out of control. Maintaining lateral stability during braking of a bob tail (tractor without its trailer) may be difficult because light braking is inefficient (35% to 40%) and hard braking may cause a spin. However, a brake proportioning valve may provide up to an 80% efficiency by applying most of the braking effort on the front wheels. When a trailer is hooked up to the tractor, the brake proportioning valve will be disconnected.

Tank trucks have the problem of movement of the liquid cargo either forward (a surge) or laterally (sloshing). Baffles and compartments help to some degree, but the hazard remains.

Pedal error occurs when the driver's foot slips off the brake pedal, gets tripped or hung up during the movement from the accelerator, or when the accelerator pedal is depressed when the driver intends to hit the brake pedal. This may be affected by pedal size, shape, location, surface covering and wear, the shoe type and size, and the knee angle.

In essence, braking error can be caused by any one of many factors, each of which requires further research for effective human error control.

(5) Mismatch errors

Putting the wrong plug into an electrical or electronic receptacle may cause immediate damage to the connection or subsequent damage to the system in terms of blown circuits or improper function at some critical future time. This has been a concern of human factors specialists for many years, and various keying or mismatch prevention configurations have been developed and widely used. For example, a keep-out feature (prevention of full insertion of the plug into a jack) is used for telephone cables (the 6-circuit RJ11 plugs), networking modular jacks (the 8-circuit RJ45 receptacle), various climate and lighting controls, and refrigerators and appliances. With increasing variations in electrical and electronic connections in automotive vehicles, this could become a problem in vehicle assembly or dealer troubleshooting of circuits. It may be a likely scenario in customized or specialized vehicles. This form of human error is clearly preventable by simple design features.

(6) Language-induced errors

A laborer opened a multiwall bag with a screwdriver. It contained caustic soda, some of which was released into the air, on to his eyes, and this caused residual visual problems. The bag fully complied with all government agency testing and package warning requirements. But there was no communication for workers who might be illiterate or could read only in a foreign language, so the human error (ripping open the package and the release of the contents into the air) was logically and legally foreseeable considering the wide marketing of the product. The injury incident could have been prevented by a design change to the package in the form of an easy-opening feature, many of which were available at low cost. Because of the high rate of illiteracy and foreign language workers and consumers, special attention should be given to the reduction of language-induced human error (see Chapter 5).

(7) Load placement errors

A driver was proceeding down a curving roadway and was surprised that his vehicle responded too quickly and too much to his attempts to steer it. He stopped and looked at the front wheels: the vehicle was tilted upward and the front wheels barely touched the pavement. He realized that he had placed too great a load in the rear bed or load carrying box of his pickup truck.

Actually, it was a load placement error. The box was centered over the rear axle, with 50% of the sprung vehicle weight over the rear axle and 50% over the front axle. The cargo load was supposed to be centered or slightly in front of the rear axle. He had committed two errors. First, the load was probably too heavy, but he recalled the sales brochures showing the vehicle carrying large rocks (a dense mass), each of the five top tie downs and the four inside-bed cargo tie-downs was rated at 700 pounds, and the vehicle had a hitch that could be used to tow boats, campers, watercraft, and motorcycles. Because of this he had assumed that the cargo box could carry a very heavy load. Second, he had used a cargo retainer or cage device to extend the bed rearward over the tailgate. Thus, an even load would be centered to the rear of the rear axle and tend to lift the front end of the machine. He did not think that using the lowered tailgate as part of the bed was an incorrect procedure because the dealer had provided the device to retain the cargo on the tailgate. In essence, the load pattern adversely affected the steering. But the cargo bed is generally intended to have versatility in carrying a variety of loads.

Some SUVs and pickup trucks have exterior roof rails, cross-bars and mounts or tie-downs to carry 100 pound loads of luggage, skis, or bicycles. Another 100 pound load could be carried on an optional rack at the rear of the vehicle. Obviously, the top load affects the vehicle's rollover propensity (raising the center of gravity) and the rear load might affect the steering dynamics (changing the weight balance of the vehicle), particularly in combination with other loads carried inside the vehicle.

Long loads, such as carpet rolls, often overhang the rear of the vehicle to the extent that it constitutes an obvious load placement error. Some drivers take pride in just how much a vehicle can carry. Some advertisements show endless groups of people exiting a vehicle or endless cargo being unloaded to dramatize the load-carrying ability of a vehicle.

Some vehicles are equipped with an air suspension system that operates on all four corners of the vehicle to attempt to distribute or level the load equally to all wheels. Some trailer hitches are also designed to distribute the load further forward on the vehicle, rather than on just a bumper hitch. Such devices do improve the handling characteristics of many vehicles.

(8) Backup errors

On a construction site, dump trucks might backup to receive a load in a slope-cut or from an excavation. They could backup to move to the proper location to dump their load, fill a void, or discharge on a conveyor. Construction sites are ever changing, with workers focused on a variety of jobs and constantly moving around the site. It was no surprise that dump trucks and other construction vehicles, with a blind-zone to the rear, would inadvertently run over a worker attending to another task. Historically, the drivers and workers were exhorted to exercise greater care and caution, but the accidents continued. The cause of the accidents was alleged to be driver error for failure to see the worker at the back of the truck, worker error for inattention in the path of the truck, and the contractor for failure to maintain a safe workplace.

Backup alarms were installed on construction equipment, then buses, then industrial equipment, and in some countries, passenger vehicles (Peters & Peters 1999a, pp. 177–183). This did little to reduce *runover* accidents where there was a blind zone at the front of the vehicle, as in large front end loaders or even forklift trucks carrying large bulky loads. The lift trucks were equipped with horns and flashing lights to warn workers of their approach whether forward, rearward, or during turning maneuvers.

Mirrors were added at the rear corners of some vehicles, for rear inboard vision close to the trucks, where the use was in residential areas and the vehicles attracted children. Some side mirrors became telescoping for a better view of what is behind and beside the vehicle. The mirrors get dirty and constantly need adjustment.

In some countries, buses were equipped with television cameras at the rear of the vehicles, TV monitors in the cab for the driver, and human assistants to control backup in crowded bus terminals. Some trucks were equipped with pressure-sensitive bars at the rear that could quickly apply the brakes to stop the vehicle if the truck hit an obstacle or loading dock.

This has grown to the availability (as standard equipment) of dual sensing systems (infrared) in the rear bumper of low-cost vehicles that can detect obstacles (such as a bicycle or children in a driveway) and alert (by beeps) the driver as to the closeness of the objects.

(9) Drinking and gambling errors

The effect of alcohol and drugs on driving performance is well known. The drunk driver is considered a menace to others on the highway. Alcohol intoxication impairs the judgment of drivers and results in erratic driving, longer reaction times, altered attitudes as to what is prudent conduct, and possible loss of consciousness. Attempts have been made to reduce this source of human error by driver education, designated driver programs, reducing blood alcohol level requirements for Driving Under the Influence

(DUI) offenses, increasing the age necessary for the sale of alcoholic beverages, and imposition of legal liability for the sale of alcohol to obviously intoxicated drivers. But this source of human error is hardly under control.

Similarly, mind-altering drugs, even at low dosages, can alter mood and accuracy of perception. Even prescription drugs may caution that they should not be used by drivers or that they may cause dizziness, drowsiness, decreased mental concentration and memory, vertigo, or even central nervous system side-effects. In other words, drugs can induce human error. Control of drug use is a humanitarian and societal objective, but there always will be drugged drivers on the road during the foreseeable future. The design of vehicles should reasonably accommodate, tolerate, or counteract that kind of conduct, assuming that no other control measures are sufficiently effective and that the drug-impaired driver constitutes a substantial risk to others.

In accident reconstruction, there are many problems relating to the intoxicated driver. Did the driver drink more than he claims? Was the driver actually impaired? An example is the truck driver who was involved in an accident that seriously injured another person. The truck driver claimed that he spent the night before the accident at a truck stop where he got a good night's sleep. However, his truck was equipped with a black box to enable dispatchers to monitor, track, and record the truck's exact location by use of the global positioning system. This record proved that the truck driver had spent the entire night before the collision drinking and gambling at a casino (confirmed by other evidence).

Some vehicles have black boxes (electronic control modules or other units) that record many parameters of vehicle operation. The accident reconstructionist may be able to determine, at the time of a collision, the throttle position, gear selection, cruise control problems, brake switch failure, vehicle speed pulsing during deceleration, and many other factors, depending on the sophistication of the system. In essence, there may be helpful objective data to substantiate or contest the claims of the vehicle drivers. Inferences of drinking, drugs, gambling, and fatigue may be better substantiated. The threat of the use of such data may have a palliative effect on those who might otherwise engage in activities that serve to increase human error potentials.

(10) Blind-side errors

For most automotive vehicles, the side opposite the driver contains a blind zone near the C-pillar. Vehicles in the adjacent lane may visually disappear as they move up to the corner area of the vehicle and attempt to pass it. The driver of the adjacent vehicle may be beyond the turn signals and unaware of an impending turn or lane change. The adjacent vehicle is not seen on rear mirrors or on the side mirrors, and direct vision is obstructed by the C-pillar or other structures.

Tractor trailers often require a large turning radius and when turning into an intersecting roadway may have to do so from a middle lane. This could trap another vehicle on the inside lane and result in that vehicle colliding with the truck. The human errors that might be claimed for the truck driver are turning from the wrong lane, failure to observe an oncoming vehicle in the adjacent lane, failure to signal the turn properly, and failure to make a safe turn. The other driver might be blamed for failure to anticipate such truck maneuvers and pay close attention to the turning truck, failure to observe the truck's rear turn signals, and failure to brake and steer his vehicle in a timely and appropriate manner to avoid a collision. To eliminate the claimed human error in such turning accidents, truck owners placed large graphic 'turn hazard' signs, with flashing lighted arrows, near the front of the trailer where it could be seen by drivers moving alongside the trailer. This was accompanied by signs at the rear of the trailer that warned other drivers that wide turns might occur. Possible side sensing devices are being developed to alert truck drivers about the approach or presence of vehicles in the adjacent lanes to better enable safe turns by drivers of oversized trucks and trailers.

Passenger vehicles have also added turn signals to the front of the vehicles. Built-in turn signals are included within the side mirrors of some vehicles in the form of flashing arrows. Other vehicles include a flashing red light adjacent to the side mirrors. These signals warn other drivers in adjoining lanes. Possible 360 degree object detection systems are being developed to provide the driver with information on the location and movement of other vehicles, including those in blind zones.

(11) Rough ground errors

Passenger car undersides frequently strike high curbs or other objects. Damage will occur if some of the vehicle structure, lines, or accessories project into the angle of approach (the angle between the ground and the leading edge of the vehicle). Skid plates may prevent damage to the understructure, if they are present.

A rutted or uneven roading surface may project high enough to damage the differential if there is insufficient axle clearance (the distance between the ground and the axle housing). Axle clearance may vary; for example from 6 to 8 inches or more. It is difficult for the driver to judge the clearance of the vehicle he is operating, but damaging the underside may be called a careless human error.

A vehicle starting down a ramp into an underground parking structure must clear the break-over angle (the maximum rise that a vehicle can clear between the front and rear wheels). The ramp break-over angle for an empty pickup truck may vary from 15 to more than 22 inches. The first awareness of a lack of clearance might be the sound of metal scraping against concrete or the stoppage of the vehicle.

Overhead clearance marks on bridges often show that some drivers are poor judges of clearance. In one case, a truck pulling a trailer loaded with a large piece of construction equipment became jammed under a bridge. The driver was familiar with the roadway and saw the bridge clearance limit sign, but he misjudged the height of the load being carried.

For off-road vehicles the terrain is often rough and uneven, with rocks and tree limbs strewn about, ditches and mudholes where least expected, and hard and soft areas on either side of the vehicle pathway. This is all part of the adventure and taking chances, particularly in very remote areas. As one vehicle manufacturer put it, there is a thrill in being on the edge during cross-country driving.

Sturdy skid plates should be used to protect the fuel tank: leakage is unsafe, especially when the fuel tank is located near the wheels and suspension systems. Protection of the frame cross-members, used to make the frame more rigid, is advisable.

An added benefit of skid plates is that they smooth out a very interrupted aerodynamic flow under the vehicle, usually with wind vortices to the side from behind the front wheels. This was discovered in France many years ago using moving platform wind tunnels.

In essence, out-of-sight protection for the underside of vehicles is necessary in urban areas, rural areas, off-road areas, and for the unimproved surfaces found in various countries. Durability may be tested in highly congested stop-and-go traffic in large cities and on the test tracks of company proving grounds, but it is also necessary for curb impacts, ramp hangups, and other clearance situations. It is necessary because humans are poor judges of vehicle clearances and poor in judging the effect on the vehicle of hitting minor obstacles.

(12) Risk-taking errors

A refuse collection truck lifted a large metal trash bin up over the cab and dumped the contents into a storage container at the rear of the cab. The driver returned the metal trash bin to the ground and started to move the lifting device to its storage area behind the cab so the truck could move on to the next pickup location. Unfortunately, the lifting device stopped halfway through the cycle. While the driver remained in his seat, the helper climbed up and on to the lift device, then the helper knew enough to start jumping vigorously to get the hydraulic system unstuck and lower the device. He had to jump hard, beside two upright sharp-edged forks, with little to grasp to steady himself. This unsafe risk-taking behavior was an intentional act, but was required to compensate for a hydraulic control problem. The lifting device gradually lowered, the hydraulic actuated cycle began again, and the truck went to the next location to pick up another load of trash. This repeated risk-taking behavior would not have occurred if the driver had known of another procedure to use on the available equipment, given the need to continue trash collections.

In the paint shop of a vehicle assembly factory, the personal protective equipment (PPE) included an air-supplied respirator, hood, and garment. It was mandatory to wear the PPE because of the presence of airborne organic solvents in the workplace. Few workers could wear the respirators for a full shift because they were so uncomfortable. Harmful exposure occurred, the injuries were recorded, and compensation payments were made. The remedy for this intentional violation of work rules was to provide sufficient exhaust ventilation so the airborne solvents would be pulled away from the workspace. Respirators were not needed, and the productivity of the workers escalated dramatically.

Protective eye glasses were required at another assembly plant, but some workers did not like the discomfort and visual aberrations. They intentionally did not use the protective eye glasses whenever they could. A cat-and-mouse game resulted between the trade union health and safety supervisor and the union workers. The problem remained uncorrected.

Drag races are one of the truest voluntary assumptions of a known risk. The risks involve high-speed single vehicle crashes, two-vehicle collisions, and police raids. The risk assumption is prolonged, starting with vehicle modifications and ending with peer applause or sympathy. While some vehicles are aftermarket specials, others have racing characteristics express or implied by the original manufacturer or accessory provider. The racing human error would not occur without the combination of drivers seeking peer approval and the need for the sensations that accompany high-risk undertakings, the availability of a roadway or track environment, and the original and modified vehicle and accessories.

(13) *Misuse and abuse*

Misuse seems to be an improper use. Abuse seems to be careless use. Both of these could be called human error. But is the error predictable? For example, there was a problem with one keyless entry system for automobiles. The keyless card became twisted and this damaged the electronic connections printed on it. The cause was male drivers who put the card in their hip pocket and then sat on it. This is certainly an abuse of the card, but changes had to be made to the card to tolerate this abuse.

Misuse of equipment is a frequent complaint after an accident occurs. But misuse is hard to prove or disprove because no one watches an operator or driver all the time. However, one model of a self-propelled aerial workplatform had been equipped with a remote monitoring system. A wireless data recorder had channels that were programmed for ignition, hydraulics, shock, cooling, oil pressure, temperature, run times, battery charging, emergency conditions, and global positioning. There are mobile cranes, forklifts, and boom lifts with data recorders that also permit evaluation of possible misuse. Some monitors have the capability of shutting down the machine if unauthorized use is detected or the rental times have expired. This can be objective data relating to human error, misuse, and abuse in the operation of construction equipment.

There may be objective evidence of *abuse* in the form of dents, scrapes, scratches, and missing parts. But, is a dent a safety problem? How is it relevant? Is it reasonable to infer that a cosmetic or appearance item is indicative of a general behavioral prediction for misuse? Does personal clothing permit inferences of possible good care, bad care, attention to the appearance of equipment, good housekeeping, good maintenance, misuse or abuse? While an accident investigator or reconstruction specialist must identify and record as much detail as possible, there is a difference between actual abuse and speculation about abuse. Of course, there may be direct observation by a reliable person or a television monitor recording to prove that the equipment was abused. If so, the question is why, so that causes and consequences may be investigated and possibly remedied.

Misuse is the improper use of a product. But what is proper or improper? 'Proper' is usually defined as what was intended by the design engineer. The operator may have believed that it was a proper use based on the information available to him. There may have been insufficient information in the form of operation instructions and warnings. Advertising materials may have over-puffed, grossly exaggerated, or misled about the capabilities of a product. Conversely, the misuse may be clearly unreasonable, an act of recklessness, an intentional disregard of prescribed procedures, or an abnormal behavioral response to some extraneous stimuli. The question is always why, what was the cause of the misuse? Was it a matter of necessity, ignorance, or thoughtful deliberation? The rule is whether or not it was a customary use or one that was predictable under the circumstances. The design engineer has the responsibility for designing a product that can prevent, minimize, or compensate for foreseeable misuse. The user has a responsibility to exercise proper care that is reasonable for persons of his peer group under the circumstances. Prevention by design is far more certain than the uncertainty of human behavior.

By way of illustration or analogy, the pharmaceutical industry has reacted to the problem of side-effects from various drugs and drug interactions. For some time, the side-effects were labeled a misuse by the prescribing physician or an abuse by the patient. Now there are detailed warnings provided in packaging inserts and in the widely used *Physicians Desk Reference* (PDR) annual book or electronic information delivery system. The side-effects, contraindications, adverse reactions, directions for use, precautions, and clinical pharmacology are described. It includes overdosage, drug dependence, adverse reactions, estimated risk of death, and drug abuse information. In other words, specific information to avoid misuse and abuse has been provided both to the physician and to the patient.

(14) Curiosity

The curiosity of children is well known. They will explore and may crawl into the trunk of a car either when the lid is left open or through an aperture between the rear seat and the trunk area. They will pretend to be a driver and attempt to steer, shift levers, push buttons, to explore, to experience, and to mimic adult driving behavior. Since they may attempt to operate power windows and become entrapped, the design prevention remedy or countermeasure may be a window locking mechanism or force-sensitive window actuators. A child's error should not have harmful consequence to the vehicle, the child, or to the parents if there is a reasonable countermeasure available.

There are child-resistant caps or covers for containers for pharmaceuticals (medications) because a child's curiosity may lead him to harm. In other words, there is a certain predictability about a child's proclivity to explore his surroundings, become curious about objects, and attempt to manipulate them with certain innocence. To some degree, this also applies to adults. When someone repeats the old saying that curiosity killed the cat, the reference is to human adult behavior. Thus, it would be advisable for the design engineer to consider what could happen from a curious adult and design appropriately to prevent such behavior from creating safety problems. Adult human error, stemming from curiosity, should not result in avoidable damage or injury.

(15) Aggressiveness

There are three basic kinds of aggressiveness that are important in vehicle safety: that of the driver, that of the vehicle, and the interaction between the driver and the vehicle. There are drivers who act as if they are competitive racing car drivers and the roadway is a race track. They rapidly accelerate, weave in and out of traffic lanes, and have a daredevil approach to driving. Traffic laws are disregarded and courtesy forgotten. Their driving mistakes or human errors are incidental to their aggressive behavior.

> It just wasn't the same after the insurance companies and the government got into the act . . . Safety regulations as well as high insurance rates had helped to cripple the muscle car market.
>
> (Anonymous, *Guide to Muscle Cars*, 1991)

A vehicle may have an aggressive appearance because it is intentionally designed to attract a particular market segment. It may be a matter of size, accessories, front-end design, and weight (mass). It may be a big vehicle driven in a traffic mix with many small and medium-sized vehicles, so size breeds a sense of superiority. The driver sits high above the others. Mass does count in a collision, under the law of conservation of energy (energy can neither be made nor destroyed, only changed in form) and the fact that kinetic energy is proportional to the square of the velocity (speed) of the vehicle. In other words, the vehicle with the greater mass and higher speed can transfer greater force to a smaller, slower vehicle. Accident reconstructionists refer to the mass ratio between vehicles (the comparative forces of moving vehicles). There is also the law of conservation of momentum (in the interaction of vehicles, the total momentum of the system remains unchanged). All of this suggests that there is more kinetic energy to be expended in deformation of vehicles, in post-crash accelerations, and in trajectory pathways. The crashworthiness of each vehicle obviously affects the collision results and light trucks do not have to meet all regulatory standards or contain all the safety features of passenger vehicles. In other words, the aggressive vehicle may or may not have safety advantage over some other vehicle. However, the belief of the driver may be quite independent of any vehicle safety advantages: he simply believes that a big aggressive-looking vehicle will intimidate the drivers of other vehicles. They will get out of his way because they fear the consequences.

There is also an interaction or combination effect between the aggressive vehicle and the aggressive driver. There is a merging of the self-image of the driver and his image of the superior vehicle. The driver may believe he has a good array of crash avoidance maneuvers, good control over the vehicle in order to execute a chosen maneuver, but very often the collision sequence indicates there is no maneuver at all to avoid the collision. The driver simply freezes in the belief that his aggressive vehicle will do well in the collision. This may be associated with the fact that many drivers of sports utility vehicles (SUVs) have been encouraged to believe that their vehicle can demolish a little vehicle in a collision.

Control over the aggressive driver is a police matter and his behavior may become manifest when least expected. In contrast, compatibility of vehicles in the traffic mix could be a design responsibility. The engineer can determine how vehicles interact or match-up in terms of driver control, visual detection of approaching vehicles, relative collision damage, and the necessary occupant protection. Design can offer some protection from the other aggressive driver. Collision warning and avoidance systems would help if designed to include the behavior of aggressive drivers. Warning buzzers or limits for higher engine rpms (acceleration and speed) and high yaw accelerations might be advantageous. Airbags of variable power for different impact velocities (accelerations), sensing and reacting to different directions of impact (vectors), and offering protection for different collision reactions (front, side, rear, and rollover) could help. This should be an area of active research and development: what to do about aggressive drivers, aggressive vehicles, and the interaction effects from both.

(16) Awareness errors

Awareness of a possible danger may be muted by satisfactory vehicle performances within the safe operation envelope. For example, a three-point farm tractor hitch may permit safe movement of pull-type implements. However, the driver may be unaware that the overhanging implement weight on the hitch itself could reduce the front axle loading of the tractor to the point where bumps, holes, rough ground, or snags could lift the front wheels off the ground causing veering or rear overturn. Low speed may be safe and higher speeds unsafe. Some bumps may not cause a problem until there is a different loading pattern in a haulage vehicle. The driver could be lulled into one state of mind, for months or years, until the final feather is dropped on the hitch and the front end of his vehicle rears up.

A new driver of a four-wheel steering vehicle may drive for some time and believe that he is aware of how the vehicle reacts to his steering input. He may be surprised that sharper turns have different response characteristics at different speeds. For electric vehicle drivability, the predictability and smoothness of powertrain response is important. Too much torque may mean too quick a vehicle response to the human input (driver actions). In essence, the vehicle should teach the driver something about its handling characteristics and not present unpleasant surprises. That is, a special design effort should be made to assure comparable responses between different vehicles, during driver transitional learning, the extremes of human vehicle handling, and the human emergency stress responses. The caveat is that the driver may not be sufficiently aware of anything unusual about a new and different vehicle or its operational limits (the performance envelope).

(17) Vibration errors

There are trucks, tractors, tanks, and other vehicles in which the driver is subject to uncomfortable vibrations of varying frequency, magnitude, duration, and complexity. The driver may complain about discomfort, fatigue, and back problems. He may attribute driver errors to the seat or vehicle vibration. Special vibration-resistant seats have been installed on many vehicles. There are front cab isolation systems (FCIS) for trucks, a vibration reducing suspension system.

On vehicle vibration in trains, hovercraft, and helicopters:

> The error rate in perceiving fine detail . . . is proportional to the product of the frequency and square root of the amplitude of the vibrating image . . . Vibration affects motor tasks, comfort, motion sickness, and headaches.
>
> (Osborne, 1986, pp. 70, 82)

(18) Negative transfer errors

If a design change is made in a vehicle, the driver's past habits may cause learning or training problems initially or the habit may reappear during a high-stress or emergency situation. This *transfer* problem may become manifest for changes in brake boosters (the distance and magnitude of force to be applied to the brake pedal for a certain braking response), for power steering and suspension changes when all four wheels are individually powered (for all-wheel drive) rather than just the rear or front wheels, and for acceleration differences between internal engine-powered vehicles and the vehicle with all electric driven wheels. The problem may be more severe if *negative transfer* occurs. One example is the human weight shifts and body movements necessary for control and stability of all-terrain vehicles (ATVs), including leaning to the outside of the turn at low speed and inside at high speeds for some vehicles. On negative transfer in ATVs:

> Unlike motorcycles and the bicycles that almost all of the population have previously learned to ride, ATVs require complex skills simply to maintain stability in many of their operations. Further there is negative transfer from the previously learned skills used in riding bicycles and motorcycles.
>
> (Karnes *et al.*, 1990, p. 13)

(19) Speed estimations

The vehicle driver may have confidence in estimating the speed of his own vehicle, but may not recognize the complexity of estimating the speed of other vehicles that are approaching from the front, side, rear, or at angles. On a two-lane highway, the driver may estimate the speed of an oncoming vehicle and move slightly away in his lane from the approaching vehicle. The lane movement (a give-room behavior) is often greatest at 3 seconds regardless of the approach speed of the other vehicle. Minor rear-end collisions are often the result of poorly estimating the relative speeds of both vehicles and the time available to brake and slow the vehicle.

> The judgment of absolute and relative speed of vehicles is a basic component of many road-user behavior tasks. A driver's estimate of speed is of great importance when approaching potential hazard sites such as curves or intersections. They also have to estimate the speed of the other vehicles and select safe clearance intervals.

> (Triggs, 1986, p. 118)

Speed estimation is necessary to preserve safe distances between the leading and following vehicles on high-speed roadways. The gap size (following or headway distance) between vehicles may be relatively small (tailgating). This may reduce see-beyond adverse information gathering and result in blind passing maneuvers. In such cases, the drivers apparently are overconfident in their ability to estimate every other vehicle's speed and their ability to evade tight driving conditions.

Speed estimation of trucks, at night, could be improved if the trucks and trailers were marked by reflective tape to aid in the visual discrimination of changes in shape, size, location, and movement. Collision avoidance systems (near obstacle detection) could provide more than simple warning signals, such as the rapidity of movement of the approaching harm.

The detection and interpretation of vehicle motion (motion analysis) involves networks of neurons in different areas of the brain's visual cortex (motion perception) and neural feedback mechanisms (information processing). This is a topic on which further research is necessary to assure the proper design of driver assistance devices. Further research is also needed in visual selective attention, controlled coupling of vision and action, and how they relate to speed estimation errors.

Speed estimation is not done in isolation; the driver must look through portions of the windshield that are bent or curved at their sides in varying amounts and factory tests are routinely performed to assess and limit visual distortion. The estimation of speeds for vehicles approaching from the rear or an adjacent lane may be hindered by C-pillar width, height, and location together with the shape of the rear side windows. In some vehicles, high seatbacks may obstruct vision to the rear either through the rear window (backlight) or by the rear-view mirror. The windshield or side windows may be tinted (colored) in a manner that adversely affects twilight or night vision. The driver may not have fully corrected vision (refraction or laser shaving) for all modes of vision. Thus, testing with a variety of drivers is essential and some manufacturers utilize eye-movement and focus data gathered by video cameras mounted in a vehicle and on the driver. Customer (driver) feedback information procedures (to the factory) are also a vital source of information.

In essence, speed estimation is very important for safe driving and continues to be an area of concern as styling changes are constantly made in vehicle design for brand identification.

(20) Assembly line errors

Assembly line workers may commit errors because they can turn their minds off during work operations because the work tasks may be extremely simple and constantly repeated for many hours, days, weeks, and months. They do not see and, sometimes, do not care about the completeness or damage on the assembly that passes in front of them. If production quotas are king, do not disturb the scheduled pace of the assembly line. Errors may be induced by parts location that require bending, twisting, and reaching motions.

In vehicle and component manufacturing operations, an ergonomics specialist may attempt to reduce musculoskeletal disorders (MSDs). There may be an effort to get the component parts off the floor, since bending and reaching into large palletized boxes or cardboard cartons (placed along the assembly line) increases the risk of shoulder and back strains. The boxes or parts can be placed on fixed-height platforms or on lift tables that can be adjusted to worker height and the available part height. They can be moved close to the point of assembly. Where there are repetitive worker tasks, job rotation may be advisable unless there is resistance to added training requirements, seniority problems, disputes over what is perceived to be easy assembly line jobs, and the possible human errors of trainees. Errors are to be expected if a worker has an aching back, tired muscles, or is inattentive.

Hand tools may be difficult to use all day, if not ergonomically designed. An emphasis may be placed on power tools, designed for a particular work task and suspended on retractable cords, and located for easy use. Videotapes of each work task or job may be taken to permit a more detailed ergonomic analysis, future use for training purposes, and possible use by medical therapists should soreness, aches, pains or MSDs result. Just the attention given to each job by representatives of management could induce positive worker perceptions that may reduce human error.

There are advocates of behavioral safety processes to reduce human error in the assembly process. These specialists may review work records to target individuals for observational analysis, then focus on whether there is compliance with the desired task or job performance, then intervene and train on the at-risk consequences of the errors, and finally attempt to effect changes in self-motivation.

During the launch of a new model vehicle, quality problems emerge all too often. Based on the plant's prior quality history of things that have gone wrong, a checklist can be developed for use in a final assembly inspection by special start-up teams of production workers and troubleshooters. There may be too much of a change in assembly procedures. An attempt should be made to standardize procedures, provide ample time to test the assembly process, and simplify the production system. During the transition to full production of a new model, appropriate use of off-the-shelf parts (carryover parts from prior generations of vehicle models) should be made to reduce the magnitude of changes occurring at one time and to better assure an uninterrupted, error-free, and smooth launch of new vehicles.

Redesign, re-engineering, and mechanical changes create greater error potential on the assembly line. Creating a new model mostly from already engineered and tested parts should involve fewer assembly problems. Restyling and trim changes should be the least troublesome. The errors that pertain to safety are most important, not the errors that could result in fit-and-finish, flushness, gap, or other cosmetic deficiencies. In essence, keep it simple, preserve assembly procedures, and provide appropriate time to make the necessary changes and become accustomed to them. On labor relations: 'The right to refuse hazardous work must be expanded to include the right to collectively refuse hazardous work and to refuse work on behalf of those at risk' (Walker, 1999, p. 47).

(21) Other errors

In the use of seatbelts (occupant restraint systems), some drivers forgot to connect the belts. Seatbelt forgetfulness was countered by a short duration audible and visual warning (belt reminders). In the future, there may be more effective countermeasures.

The upper torso (shoulder) belt might be anchored on the B-pillar, near the belt line (under the side window), near the floor, or elsewhere. Its pathway might be directed, by various belt loops or guides, toward the occupant's shoulder. A long belt introduces stretch, tangle, and belt material degradation problems. If the seat back were tilted rearward, reclined, or sometimes positioned rearward, the belt might separate from the shoulder and provide little restraint. In some cases, this permitted unrestrained ramping of the occupant to the rear of the vehicle in rear-end collisions. Such problems were solved by locating the upper end of the shoulder belt in the seat back, so the belt stays in proper position over the shoulder as the seat moves or is reclined. The belts-in-seat devices (retractor in the seat back) overcome the human adjustment problem.

The virtually uncontrollable movement of children in the rear occupant space and numerous injuries resulted in child-safe door locking systems, pressure-sensitive window closing, and open door warning devices. Forgetfulness about closing windows and locking doors led to anti-theft remote control devices and wireless coded ignition keys.

Headlight errors included when, exactly, to turn on the lights at dusk and to remember to dim the headlights for oncoming traffic. The higher headlight location on SUVs and higher intensity headlights often produce glare for oncoming vehicles. One remedy is for the headlights to be turned on automatically according to the ambient light conditions or put into low beam automatically by sensing approaching head-lights. Other vehicles have daylight running headlights that are on all the time, with the side benefit of increasing the vehicle's conspicuousness (likelihood of being seen by other drivers).

Vehicles carry various objects. There is some carelessness about such objects being placed in or on the vehicle. One sales brochure shows large and heavy wood, plastic, and metal objects being carried in the interior occupant space, a flying object safety hazard in a collision. Tires are no longer carried in the occupant compartment regardless of how secured, since they may not be fixed or locked in place as intended. Groceries placed in the trunk now have cargo nets to prevent their movement during rapid acceleration or deceleration of the vehicle. Every object in the interior or exterior should be properly restrained to prevent damage from movement or to prevent flying missiles from endangering the vehicle's occupants or the vehicles and pedestrians in close proximity to the vehicle.

During manufacturing operations at the vehicle assembler or supplier, there have been a variety of human error problems and remedies. Metalforming presses, such as punch presses and press brakes, were traditionally a source of injury because the work often required hands between the open dies, to locate and then remove the metal piece being worked upon. A quickly descending ram could injure a hand before it could be removed from the die space. A variety of safeguards were attempted over a 100-year period. Today, light curtains and presence-sensing devices prevent the press from cycling while hands are in the die space. The most common safety device is to require two palm buttons to be depressed before the press can be actuated or to use the dual buttons for actuation since both hands are out of harm's way. They may be supplemented by barrier guards and special feeding mechanisms. The human error of reaching into a danger zone at the wrong time now has an effective remedy.

A machine tool or a vehicle assembly line can be stopped, if there is a physical problem or to minimize personal injury risks, by the use of hanging pull stops, horizontal cable shut-offs, push-bar stops, or workstation emergency stop buttons. To prevent surprises, before the machine or line starts up again, a warning sound and light may signal workers to get out of the way or to prepare to start work again. Similarly, on large machines a worker may be in a blind zone, particularly during cleaning and maintenance operations. A lock-out tag-out procedure or interlocked start buttons (one inside the blind zone and one at the operator's console) may be used. For each kind of machine, there may be a trade standard that suggests appropriate safeguards to compensate for human error, or the machine may need special safeguards if part of a larger grouping of machine tools, conveyors, and assembly work stations.

Other human error problems are presented throughout this book. For example, see missteps on toeholds and running boards in Chapter 12 section (c), and neglect in Chapter 12 section (b). Also see distraction errors in Chapter 7. There are many less frequently occurring human errors. It is important to remember that as new devices and accessories are developed, history demonstrates there will be new human error problems to be resolved.

(d) Acceptable error

Human error may be pervasive in certain workstations on a vehicle assembly line. It may be from the tedium of simple repetitive motions during a 10-hour shift involving the performance of more than 500 identical motions or tasks. It may be from relief workers who ignore ergonomic assist devices. The error may be controlled only to some limited degree, permissibly more error if it is unimportant and less if it really matters. The cost of human error reduction may require considerable explanation to management and passage of time before remedies can be instituted. The golden rule is that human error is the root cause of almost every product discrepancy and defect. The objective is to eliminate human error in order to eliminate discrepancies.

An acceptable error rate may be calculated, for guidance and audit purposes, for a specific event, based on the unique circumstances and available risk assessments (see Table 4.1).

A great deal of effort might be expended to prevent (reduce) unwanted human error (deviation from what is desired or necessary in terms of human performance). For safety-critical errors this might be justified, but the cost often increases disproportionately as the final increments of improvement are attacked (prospectively or remedially). Fortunately, most human errors do not have immediate serious consequences and may be capable of quick detection and correction before there are any adverse complications. It may be more effective to have a means to audit and correct probable errors, as they

Table 4.1 Human error (non-critical): illustrative criteria that may be adapted to a particular application

Level of error	Level of control
0–5%	Uncorrectable or residual
5–10%	Partial control
10–20%	Poor control
20–40%	Inadequate control
>40%	Out of control

occur, by computer software, machine devices, or by employing one or more of the safety concepts in Chapter 2.

(e) Preventive measures

The control of human error depends on the recognition that it is not a matter of achieving an esthetically pleasing product, applying common-sense logic in the vehicle assembly process, or an instantaneous insight into the cause of human errors. Instead of subjective intuitive evaluations, something more is needed.

(1) Proactive plans

Early discovery and resolution of human error problems is certainly better and easier than late discovery. While the primary emphasis should be proactive, a secondary effort is necessary to monitor for human error and to provide quick reaction when unanticipated problems arise. This requires the application of relevant techniques using appropriate knowledge early in the design process and life-cycle of the product.

Probably the first step in organizing a systematic and useful human error control program is to compile a list of troublesome errors that have occurred in the past or in ongoing processes, products, systems, and services. Starting to collect real human error data can help to initiate the process and the ability to realistically highlight potential future problems can enable factual comparisons with new designs, and can serve as the fodder for design reviews. The human error list should include how the problem became manifest and its criticality, the specific underlying human error identified and the representative causal behaviors involved, and the corrective action undertaken and its effectiveness.

(2) Positive perspective

Trade union workers must understand the company's need for productivity, i.e. the other party's perspective. Similarly, a company engineer must understand the needs and behaviors of workers, users, and consumers, i.e. the workers' and the customer requirement perspective. This dual perspective should be positive in nature, since pure self-interest, tokenism, and negative feelings will obstruct the necessary understanding and creative solutions to problems. Most proactive techniques involve uninhibited prediction of failures in order to recognize what preventive remedies may be necessary. A positive attitude facilitates the search for possible human error and its eradication.

(3) Open causation

Human error control requires the identification of some sort of error, mistake, or failure. This serves only as a premise or basis for the initiation of a search to determine causation, i.e. if the fault is unsafe speed this is a premise, but control is based on how and why speeding occurred. The search should be open to any and all causal factors, since several substantial factors may be involved in the causation of a single human error. For example, speeding may be an obvious initial cause of an injury-producing collision. But the risk-taking behavior may have motivational factors unique to the driver; the vehicle speed may have something to do with the vehicle acceleration, handling, visibility, and road feel (feedback) characteristics; and vehicle crashworthiness may be a prime preventive factor in the resulting injury.

(4) Empirical data

Speculative information on human behavior may have little value. The objective is to obtain empirically based knowledge concerning real-life discrete behaviors that are directly relevant to what is needed. In other words, is there justifiable reliance when using human error information? The source of such data might be from unobtrusive in-field observation, videotaped work activities, human performance on mockups and prototypes, and historical experience with similar tasks and products. Some companies employ covert surveillance and shadowing of customers or their counterparts, but this is an obvious invasion of privacy that is not needed to collect valid, practical, and useful observational data.

(5) Responsible person

The designation of a responsible person or group is necessary for the accumulation of specialized information and effective implementation of a human error control program. Otherwise, if everyone is responsible, no one will really exercise that secondary or supplemental responsibility in an effective manner. The responsible person should have special training relevant to the tasks involved. For example, one company assigned an assembly worker the task, gave him some leaflets to store at his workplace, and still required him to do his assembly tasks full-time. He did not have either the knowledge or the time to do the job. Another company assigned the task to a union health and safety worker who had full-time on-the-floor productivity responsibilities. He admitted that they did some good, but could do much more. In still another company, the responsibility was given to an engineering manager who enjoyed the added responsibility and attempted to use his own common sense in identifying and resolving human error problems. But common sense is not enough.

(6) Localization

Customizing, private brand conversions, rebuilding, and other modifications of a product are commonplace. Some may be adaptations to a local population, culture, or demographic population. Human error control should include both the generic product and subsequent attempts at localization, remanufacture, or adaptation to special needs. Localization is the attempt to tailor a product to a local market: for example, raising a sedan's roof line (headliner) to accommodate the mandatory headdress of a particular subgroup, raising the speed of a stock car, or adding trim changes and increasing the horsepower of a vehicle intended to appeal to the youth market.

(7) Validation

This term is often used to describe the proofing of design changes. Similarly, the human error controls should be verified as to their effectiveness. If they consist of post-marketing processes, they should be periodically exercised if that is appropriate, for example recall or notification efforts. The recall plan should be reviewed for its human error potential. The validation process itself is not immune from human error; that is one of the reasons for exercising or simulating the process. Testing and retesting an interactive process may be necessary in order to achieve an appropriate level of human error control.

(8) Consequences

Human error can occur in products, processes, services, and engineered systems. Human error control may substantially reduce the risk of occurrence of the error. But it should also examine, *protect against*, and control the *consequences* of human errors of commission and omission. The consequences may be worse than the human error problem being corrected.

(9) Coordination of efforts

It is usually more efficient and effective for those performing human error control activities to work in cooperation with others performing related systematic analyses and evaluations. This includes reliability engineers who perform failure mode and effect analysis, system safety engineers performing fault tree analysis, system engineers performing independent design reviews, test engineers performing proving ground studies, and warranty claims personnel. Unfortunately, cross-border cooperation is often lacking in organizational structures and engineering disciplines. Open borders and free exchange of information can have dramatic results.

(10) Conceptual caveats

It may be productive, for purposes of analysis, to segregate error as follows:

* *human error* includes the understandable and excusable human mistakes
* *human failure* includes the unfortunate but non-excusable human performance shortcomings
* *human fault* includes the blameworthy, the punishable, and damage-producing events.

Remember these final reminders on human error:

* attempt to control the error, not the person
* people vary; accommodate them

- consider the context and dynamics, not the static and sterile
- observe and analyze intrinsic human characteristics, then find a means to compensate for them
- passive compliance alone may inhibit leadership in the search for fundamental and real solutions
- a situation or condition may be deemed unsafe or unacceptable depending on customer expectations, the ease of correctability, or the unacceptability to others of the errors.

If left to themselves, machines do not stay adjusted, components wear out, and managers and operators forget, miscommunicate, and change jobs. Thus a stable stationary state is an unnatural one and its achievement requires a hard and continuous fight.

(Black Nembhard, 2001)

5 Risk communication

(a) Introduction

The communication of risk is everywhere in the automotive industry; sometimes it is excellent, occasionally it is poor, and often it is overlooked as an integral part of vehicle design. Risk communication is implicit in vehicle turn signals and brake lights, rear reflectors and side lights, horns and backup alarms, battery and fuel warnings, tread-wear indicators and oil-pressure lights, airbag and collision avoidance warnings, tipover and tilt alarms, low tire pressure and navigator warnings, jack and bead seat warnings, recall and retrofit notices, and the warnings contained in service and owner's manuals and safety bulletins. There may be printed, visual, tactile, vehicle vibration, or auditory warnings.

The roadway environment in which a vehicle may operate has many warnings such as construction zone and pedestrian crossing signs, yield and wrong-way signs, reflective pavement markers and plastic bumps or rumble strips, railroad–highway grade crossing cross-buck signs and barrier gates, and traffic safety cones and guide posts.

The dealer is aware of 'right to know' stations containing material safety data sheets for its employees and containers carrying chemical hazard labels. There are danger, caution, and notice signs. The trucker and fleet manager are aware of the flammable, corrosive, combustible, and gasoline placards required for their trucks. In factories, there are signs indicating that forklifts are prohibited until truck wheels are chocked. Similarly, design engineers are aware that there may be cautions in design manuals, drawings, specifications, requirements, bill of materials, or design analysis documents. In essence, risk communication labels, devices, and documents abound everywhere.

Risk communication is the process of presenting information concerning hazards and their risks. It is generally limited to possible unacceptable risks. It is directed to all those who might take informed avoidance behavior, either preventive or corrective. In essence, it is giving others a fair opportunity to consent or make an informed choice as to the risks of personal injury or property damage. The risk messages or signals should be appropriately targeted, communicated, understood and comprehended. The protective behavior or precautions that should be undertaken to minimize the risks are an important element of risk communication.

Risk communication may seem like a simple task, but this is where the danger lurks. An engineer may create a warning that results in no behavioral change, so it is ineffective except as to cosmetic compliance. In some situations an ill-conceived warning may have reverse effects, so it increases the risks. A more sophisticated approach will be followed in this chapter.

(b) Imagery

There are forms of communication that involve a steady flow of information to establish credibility for future messages. Automobile advertising helps to sell vehicles, but it also establishes a financial connection with the print, radio, and television media. One newspaper publisher told one of the authors of this book that automobile advertising constituted most of the profit in his operations. Financial interest provides media access and credibility when needed. Lobbying by automotive industry pubic relations specialists also means campaign contributions and a resultant financial connection with politicians. It also provides derivative credibility with government regulators. This storehouse of access and credibility may be available for future spin activities relating to risk communication.

Spin is the embellishment and extension of facts, the rationalization and advocacy of arguments, and an attempt to create beliefs and opinions. It goes beyond simple excuses and may resemble the art of propaganda. It is called *white spin* if it is favorable and serves to create a good image. It is *dark spin* if it serves to create an unfavorable image. It may be *clear spin* for transparent or neutral communications that can be fact-checked for accuracy and serve as image-building, source identification for media access, and the development of possible friendships.

An example of dark spin, to create an unfavorable image, occurs when a credible source experiences success in attaching a label to a whistleblower, tattletale, or even a worker giving unbiased disclosures if the workers have not established their own credibility. Simply calling the worker a 'disgruntled employee' or someone seeking 'retaliation' can foster perceptions of untruthfulness, combat altruistic impressions, and create negative imagery in the predisposed. It should be clearly understood that some spin is involved in nearly every communication, but *managed spin* is a highly focused, clearly defined, message-oriented, and target audience directed effort.

In terms of risk communication, dark spin could serve to cloud the issues, create uncertainties about specific risk, engender questions about methodologies, and delay avoidance or corrective behavior. Thus, dark spin could easily result in avoidable injury and preventable damage. In essence, where there are hazards that present appreciable risk, any type of spin should be avoided. Let the facts speak for themselves, and the risk communication be as effective as possible under the circumstances.

In essence, spin should be recognized as helping to 'set the stage' by developing the imagery that determines the believability of opinions regarding risk. It facilitates or hinders risk communication.

(c) Urgency

During the early stages of design, the process of hazard identification and risk assessment may be highly subjective and somewhat speculative. The risk communication may be only question-oriented, because of the uncertainties. As time progresses, system safety analyses become more quantitative, objective, and supported by logic, data, and test results. The risk communication may become much more specific and remedy-seeking. As design drawings become ever more detailed, the possible hazards and risks need to be openly placed on the drawings so they cannot be ignored, forgotten, and overlooked. Assembly drawings should contain cautions where safety might be compromised by cumulative dimensional tolerance buildup, materials substitution, or lack of quality inspection.

As the risk becomes more specific and certain, it assumes a greater sense of urgency. Warnings should reflect that urgency. For example, cigarette (tobacco) warnings were, at first, informational in character, stating 'Underage sale prohibited' and 'Surgeon General's warning: contains carbon monoxide'. The hazards, i.e. addiction and lung cancer, among others, were not revealed. The urgency level that was communicated was low. The warnings had a minor effect on smoking behavior. More graphic pictorial illustrations (sometimes a picture of a victim over half of the package) seemed to make more smokers think about the consequences of smoking. Special taxes to increase the price of cigarettes began to serve as a disincentive. This merely illustrates the general rule that warnings often need reinforcement (repetition in various forms or enforcement with penalties). It also illustrates the fact that the greater the urgency and certainty of injury or damage, the more graphically urgent and attention-getting should be the warning or risk communication.

(d) Magnitude of the risk

The risk of a particular hazard may be classified *quantitatively*, such as the number of specified injuries expected over the service life of an item, for equipment produced in a stated quantity, and used in a defined market or environment. The risk may be defined *qualitatively* as high, medium, low, or negligible. Estimating the magnitude of the risk is important in terms of engineering management, but also in terms of what is represented or conveyed to the purchaser, user, or consumer. The risk magnitude, however stated, tends to establish consumer expectations and perceptions of the need for precautions.

A review was made of the owner's manuals for two similar passenger vehicles. Both utilized highlighted boxes for warnings, which were as follows:

Company A
Warning: Could result in an accident or bodily injury.
Caution: Could result in damage to the vehicle.

Company B
Warning: Uses the word 'Caution' to mean a warning of something that could cause injury.
Notice: Could result in damage to the vehicle.

The data in Table 5.1 shows that one brand of passenger vehicle represented that it had 169% of the other brand's injury-causing hazards and 246% of the other brand's damage-causing hazards. That brand also required 72% more pages for the instructions, even though both booklets were a 5 × 8 inch size and used similar page layouts and type size.

Table 5.1 Owner's manual data

	Brand A	Brand B
Total pages	210	362
Warnings	58	0
Cautions	28	98
Notices	0	69
Total	86	167

Was Brand B more honest in reporting nearly twice as many hazards in its owner's manual? If so, why did it downgrade the definitions so that the owner's manual contained nothing with the signal word 'warning'? Is this confusing or misleading for the consumer who reads the owner's manual? Could it be based on a difference of opinion as to the magnitude of the risk?

System safety engineers have more clearly established guidelines for estimating the magnitude of the risk, as reflected in the articles that have appeared in the *Journal of System Safety*, published by the System Safety Society. Harmonization was required for management decisions, under the provisions of MIL-STD-882D, *Standard Practice for System Safety* (available at http://www.google.com/search?q= cache: GfSvnpZO654C: nasdac.faa.gov/RISK/SSHandbook/app_h_1200.PDF+MIL-STD-882D&hl=en). The risk is generally classified as to frequency (improbable, remote, occasional, probable, or frequent) and by severity (negligible, marginal, critical, or catastrophic). Then, risk may be categorized as low, medium, serious, or high based on probability estimates as to costs (however defined). Such risk estimates may be affected by biases such as the risk aversion or risk tolerance of the analyst, his supervisory peers, or management.

(e) Readability

There are many methods that can be used to determine the readability of written text used in communication. The objective is to assure that the words or phrases can be read by the target audience. If there are automobile owners and drivers who cannot read the information presented, some other means of communication should be used. This is particularly needed for those literate in a language other than those languages available in the manual.

It should be apparent that the ability to *read* warnings, cautions, safety notices, directions, and instructions is only the first criterion. The next question is whether what is read is actually understood. This is the *meaning* of the words and phrases standing alone. For example, the driver may read and interpret the words to mean that he understands that the reclining seat is for personal comfort. The third and most important question deals with comprehension. Unless the full intent of a safety message is comprehended, it will be quickly forgotten as having no real importance. For example, that the reclining seat should be in the upright position while driving *because* of the likelihood of ramping and personal injury during a rear collision. Another example is that the word 'safe' can be read without any meaning attached, it can be understood as being safe in the passive sense of a general meaning, or it can be comprehended as actively alerting a person as to how to avoid a specific injury. Still another example: a person may read out loud the word 'carcinogen' but be unable to explain what it means. Some may understand that it means cancer-producing without being able to explain how, why, or what to do about it. Those who comprehend the specialized meaning can discuss causation, protection, and avoidance behavior.

This suggests why complex words and obscure symbols should be avoided in warnings and instructions meant for the general public. Fudge words and conditional statements, to avoid 'scaring' the reader, should be avoided because they confuse and defeat the direct message being communicated.

Warnings and instructions that are not practical, reasonable, or that have high personal compliance cost will be ignored. Do they conform to the custom and practice of the peer group? Are they compatible with cultural variations? Do they conform to community standards?

Are the cautions and warnings tailored or customized to the specific hazard in the applicable model vehicle or are they far too general?

Table 5.2 shows the readability data for automotive vehicle manuals intended for vehicle post-sale use. The owner's manuals and service manuals vary in reading ease. Some are 'standard' or at the reading level of the average person, meaning that half the population would have difficulty reading them. Others are 'fairly easy', but may still require some high-school education. The repair manual was 'fairly difficult' and one assessment suggests some college training or higher education would be necessary.

As can be seen, readability varies from manual to manual. It also varies from section to section within each manual. Different scoring systems provide different results. However, any risk communication document can be assessed and, perhaps, rewritten to improve readability. It takes a failure to understand just one key element in a warning to render it ineffective.

Just about anyone in the general population can drive a vehicle, which is a tribute to those who have made the process so simple and intuitive. But all drivers are not equal. Perhaps some lack the information contained in the manuals, and, in particular, risk communication. The bad driver, the poor vehicle service at a dealer, and the difficulties of some auto-repair specialists may be associated with information deprivation.

(f) Specific examples

The warning in Figure 5.1 is placed in a protected position inside the door area. Thus, it seems to meet the basic warning design criteria of durability and placement when and where needed. The question is whether it will motivate the driver to avoid 'high-performance tires', a key objective of the message. The terms 'rollover' and 'serious injury' serve both as incentives for compliance and the necessary information as to specific consequences.

Cautions can be applied in unusual locations, such as the round bar of a moveable tailgate cage or bed extender on a pickup truck as shown in Figure 5.2. Again, the

Table 5.2 Readability

Document	Flesch-Kinaid		Coleman-Liau grade level	Bormuth grade level
	Grade level	Flesch reading ease		
Owner's Manual				
Company A	7.14	Standard	13.89	10.50
Company B	5.83	Fairly easy	8.39	8.90
Service manual				
Light truck	6.87	Standard	10.65	9.50
Body	7.40	Fairly easy	11.91	10.30
Repair manual				
Light truck	9.36	Fairly difficult	15.90	11.10

A Flesch grade level of 7 indicates that the writing can be understood by an 'average' reader who has completed seven years of education in the United States.

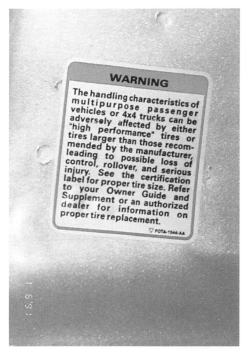

Figure 5.1 Example of warning in door area.

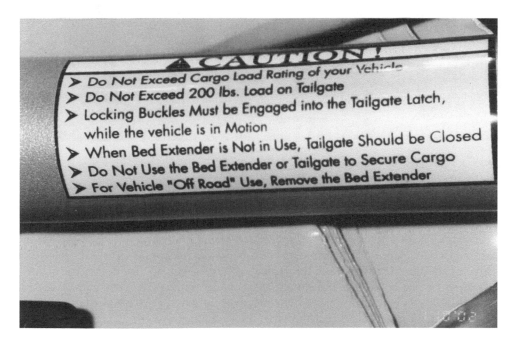

Figure 5.2 Example of warning on bed extender bar.

warning is placed where needed. The message is that the additional load that can be carried is 200 pounds on the tailgate, as contrasted with other vehicles that limit the load to 100 pounds. Such loads are leveraged and tend to lift the front wheels causing oversteering and other handling problems, particularly if the load is not evenly distributed over the entire bed of the truck. Such bed extenders are both factory options and aftermarket accessories.

This warning uses the signal word 'caution', suggesting to some that the risk is confined to property damage. The off-road prohibition may suggest that the rotation axis and fixation pin may not be adequate. The warning design principle to be remembered is that the warning should not have 'side-effects' or unintended meanings.

Warnings can be in more than one language, which helps in border areas or in pockets of foreign-language residents, as shown in Figure 5.3. The message describes the consequences (killed or injured) of placing children in the front seat of passenger vehicles. International symbols can be used. Instructions are given not to remove the warning label, but the same warning, shown in Figure 5.4, was hung on a piece of plastic from the glove compartment, so it could be easily removed despite the instructions. A message lost is a failure to warn. Such a warning is easily tested to determine whether any behavior modification is achieved by use of the warning.

Figure 5.3 Example of bilingual warning.

The belt buckle shown in Figure 5.5 contains risk communication information labeled 'important'. The engraved metal is certainly a durable form of warning, but how many people would look at the back of the buckle and actually see the warning? Whether the warning design criterion of conspicuousness is met is questionable. It does refer to a source of additional information: the vehicle owner's manual.

A warning can be simple and direct, as shown in Figure 5.6. The signal word is 'caution', although drilling into a high-voltage line may cause more than property damage. Automotive vehicle facilities often have temporary warning signs that do not meet trade standards or regulatory agency requirements. Conversely, too much information can be presented, with instructions mixed with warnings (Figure 5.7).

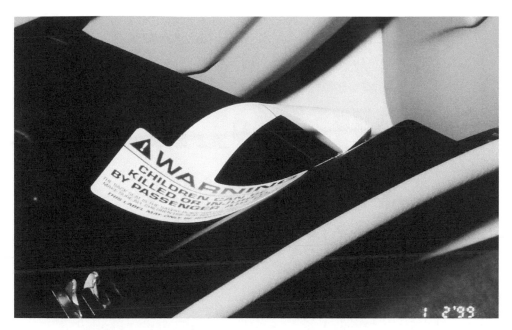

Figure 5.4 Example of easily removable warning.

Figure 5.5 Example of inconspicuous warning.

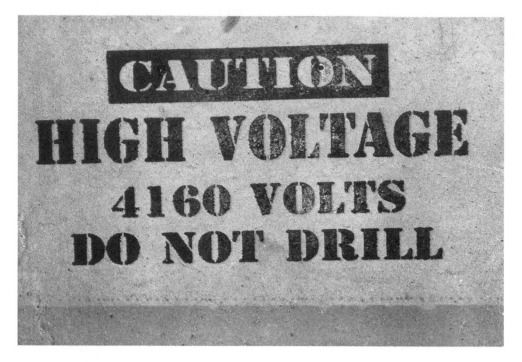

Figure 5.6 Example of simple warning.

Figure 5.7 Example of warning with too much information.

Internationally accepted pictorials, supplemented with English-language words, may solve many risk communication problems. This is shown in Figure 5.8 for a battery warning. A pictorial on a seatbelt is shown in Figure 5.9. The pictorial indicates that it is important to consult the owner's manual about children's seats.

Figure 5.8 Example of warning with pictorials and words.

Figure 5.9 Example of warning with pictorials only.

Pictorials prominently displayed can quickly communicate risks such as hand injuries from belts, as shown in Figure 5.10. When the risk has been lowered by battery design and special attachment points, a clear and simple direction may suffice, as shown in Figure 5.11.

Figure 5.10 Example of well-placed pictorial warning.

Figure 5.11 Example of low-risk warning.

The design guidelines presented in this chapter apply to paper warnings, placards, signs, inserts, folded labels, hang-tag booklets with resealable adhesive strips, and instructions on a string. They also apply to elapsed-time warning devices, out-of-tolerance warning devices, or hydraulic fault indicators. In essence, the design principles are generic. They may be used for emergency purposes, as reminders, or simply as secondary advisories. The effect may be positive, neutral, or adverse in terms of behavior modification. For further design details see Peters and Peters (1999a).

6 Universal design

(a) Introduction

This topic is discussed in Chapter 2 as a basic vehicle safety concept. Its importance necessitates a more detailed examination with more examples. The basic concept is universal design for people, i.e. design in terms of the broad range of users, not just for the so-called average person or some small group of people that excludes the extremes of a normal population distribution. The design requirements should accommodate all those who represent the foreseeable, predictable, and anticipated user population.

The universal design concept has been gradually broadened so that vehicle designs should be compatible with all users, under all environmental conditions, for all anticipated uses, and for all predictable modifications. This may seem logical and common sense, but traditional design does not typically encompass such broad design responsibilities. Generally, it is asserted that it is the user who must adapt to the machine and learn how to use it correctly and responsibly (the safe use doctrine), rather than having a universal design to accommodate and safeguard the user.

(b) The theory

The current universal design concept helps to ensure that a far greater proportion of people (the target population) will be satisfied with a product (system) and not be harmed by it. It is a necessary part of system safety where risks are systematically analyzed and hazard prevention is a goal. Universal design requires greater consideration of human factors, including procedural human error and predictable misuse. It involves a change in perspective, since evaluations, in general, are made in terms of world trade criteria and global legal requirements. For example, the European Commission has adopted the 'precautionary principle' in risk assessments. This defines a danger as when there are reasonable grounds for concern about potentially dangerous effects on humans, the environment, animals, and plants, particularly in situations where scientific evidence is insufficient, inconclusive, or uncertain.

In the United States, a lower level of protection has resulted from the decisions of the Supreme Court on what constitutes reliable scientific evidence. However, other case law, in the United States and around the world, has expanded on what is foreseeable (possibly needing precautions) and the standard of due care (reasonably practicable preventive action). The somewhat ambiguous and varying concepts as to what is or is not an acceptable risk or 'safe' suggests that there should be a reasonably prudent and factual basis for the overall design of a product or system. This should include universal design principles which utilize objective methods for determining foreseeability and the actual standard of care to be exercised.

For further information on risk assessment see Chapter 3. Universal design is necessary for favorable risk assessments.

(c) A reasonable standard of care

To avoid personal injury or property damage, the legal term 'reasonably foreseeable', in terms of the consumer or user, suggests that there be some awareness and specific knowledge of the broad array of people who might use or interact with a product, process, or service. Once the 'people array' has been identified, ordinary prudence or reasonable care (avoidance of negligence) suggests that appropriate measures be taken to prevent any unreasonable or easily avoidable risks of harm to that array. But what is the 'standard of care' in regard to people, injuries, equipment, and design characteristics as the term 'foreseeable' and 'preventable harm' have become better defined? *The more inclusive terminology for modern foreseeability may be called universal design.* It is closely related to the legal, social, political, and personal objective (criterion) of achieving a higher individual quality of life for all persons.

(d) Representative users

In the early history of design, the engineer utilized himself to determine whether a product had a good anthropometric 'fit'. The designer subjectively determined whether the product was comfortable and easy to use, whether it could adequately perform the designated task to be performed, and how it might be assembled or disassembled, serviced or repaired, and sold or recycled. It was considered sufficient that the engineer could perform the necessary procedure or operations, read and understand the instructions, and achieve the desired result. Obviously, the customer or user might be quite different in physical size, mental capability, training, familiarity with the product, and customary practices or expectations.

This designer-oriented approach to products, processes, and services reflected the technical familiarity and superiority of the producer of goods and services. Similarly, a one-size-fits-all philosophy was thought necessary for economic mass production, simplified distribution, and rapid sale. The consumer simply should exercise prudence and caution, and be aware of potential problems in the unique circumstances of the product application. The consumer almost always had the right *not* to buy the product, process, or services.

Gradually, the concept of the 'average person' (not the designer) as the user model emerged. It was typically the 50th percentile male exemplar. Later, the average female was included. If children were involved, it was the average child or average for each age group. Of course, few of the general population were 'average'. If the variance was large, a majority of people might be excluded from consideration. Of primary consideration was physical size. Obviously, differences existed in other physical characteristics, aptitudes, capabilities, personality, attitude, reading level, intelligence, motivation, and interest. Recently, as tests were conducted with a larger (cross-sectional) group of people on one or more variables, it was assumed that they were also representative of a broad array of other human characteristics or attributes.

The US National Highway Traffic Safety Administration, in its New Car Assessment Program, uses 'dummies' in its crash tests. But they are all models of a 5 foot 8-inch male weighing 171 pounds, i.e. the average man. Recently (Stoffer, 2000), the government agency requested funds to use a 4-foot 11-inch, 108-pound, female dummy, i.e. the average female. Thus, all of the past crash data may have questionable applicability to a broad range of people.

In some circumstances, appropriate design consideration may not have been given to the young and old. There may be specific regulations pertaining to the disabled, but even if not directly applicable, do such laws imply that a more universal design should be considered, in general?

At first, the group was defined as including the 5th and 95th percentile persons (the middle 90% of a given population, which excluded consideration of the 10% not covered). This changed to the 2.5th percentile and to the 97.5th percentile (which roughly included ± 2 standard deviations, but still excluded 5%). Such concepts were broadened to include the 2.5th percentile female to the 97.5th percentile male. Then the 6 sigma (±3 standard deviations) approach attempted to include 99.75% of each population group. The reason for considering a full range of foreseeable users was to prevent misfits and 'human errors' likely to cause injury, liability problems, performance failures, contract violations, and system inadequacies.

(e) Idiosyncratic risks

Some products, such as hair dye, may not be safe for a small number of consumers, but what exactly constitutes a small number when the adverse reaction may be moderately severe or life-threatening? Is it associated with a genetic predisposition to develop allergies, excessive overuse or misuse, a personal idiosyncratic reaction, a sophisticated user who ignores warning symptoms, or just a normal side-effect? The hair dye may have a warning requiring a consumer to spot test, with a time delay, and a reasonable procedure (instructions). Latex products, such as gloves and enema tips, may have trade association warnings or use substitute materials. Other illustrative products include various forms of breast and penile implants, pedicle screws, and diethylstilbestrol (DES). The basic question is what constitutes a representative consumer group for design objectives (specifications), manufacturing (quality control) capability, and testing for safety. How broad a range of people should be included for a particular type of product? Should a representative population be identified in all product risk assessments?

Proper selection of appropriate operators (users) in industry has become an increasingly futile proposition for technically complex equipment, i.e. selecting the people who could fit the machine. Thus, it seems better to have a universal design to accommodate nearly all individuals, cultures, and demands. Certainly, people excluded from design consideration and exposed to product harm should be warned and instructed (Peters and Peters, 1999a).

(f) Adjustability

Products may be designed for a broader range of people in many ways. Automobile seats have become adjustable as to horizontal distance from the steering wheel and in height for better vision of the roadway. Recently, adjustable brakes and accelerator pedals have been incorporated so as to better position a broader range of drivers. In other words, mass-produced products need not be customized, they can have built-in adjustment features.

Some companies use focus groups to better tailor the product to the market. Some perform laboratory testing, small-town testing, or apply field data in assessing preproduction products. However, such testing is frequently very limited in scope and character and is sometimes ignored in the rush to get the new product into the marketplace.

Currently, most products are actually tested by 'field analysis' (marketplace experience). This often supplements limited company evaluation or compliance testing. Instead of just 'sell and forget', more attention is being given to purchaser complaints and legal claims. Warnings may be issued, silent recalls instituted, modification kits issued, and accident investigations conducted (for both liability defense and future product improvement). Basically, there should be something similar to a real-world test with subsequent efforts to control risks and costs. Reliance on post-sale testing is not proactive or entirely in the public interest, particularly if you, inadvertently, are one of the test guinea pigs used to discover and remedy problems.

In some industries, attempts are being made to standardize parts design systems, conduct joint development projects, produce common subassemblies for purchase by different companies, and offer 'human tested' safety devices for all products in the industry. Such standardization efforts not only reduce costs, but generally provide greater assurance that all design variables have been appropriately considered and tested before a product is introduced into the marketplace. However, the company that accepts or purchases a standardized component or subassembly for inclusion in its product should exercise oversight to assure that universal design was accomplished by using an appropriate standard of care.

What all of this suggests for future product design is that market segments must be identified, products adapted and tested for that market segment, and flexible on-demand manufacturing processes utilized for products targeted at local demand. Product labeling and instructions should be appropriate to the product configuration for a particular market segment.

(g) Injury levels

At present, a certain frequency or percentage of injuries may be tolerated in terms of risk management, either by companies or in government standards. An illustration is the head injury criterion (HIC) found in company design manuals and government regulatory documents and used for automobile crash testing. The HIC of 1000 is used as a pass–fail safety criterion relating to life-threatening brain injuries, yet that level is at the 16% to 20% injury level. What is an acceptable risk level for the producer or government entity may not be acceptable to the consumer or user. How should the difference be resolved?

Someone exposed to a chemical may believe that there is no risk if the exposure is below a permissible exposure limit (PEL), a threshold limit value (TLV), or a maximum allowable concentration (MAC). However, these may be rather arbitrary limits with a known or some unknown risk of harm. Even a no significant risk (NSR) or a non-biological effect level (NBE) may be based on one excess cancer (or more) base level criterion, with a specified relative risk. The risk also varies as to whether the exposure is workplace or residential, chronic or episodic, and relatively stable or varying. The risk is, therefore, one that is acceptable or permissible for the producer. An individual may have a different tolerance, susceptibility, or sensitivity. Since good toxicological data are not available for many commonly used chemicals, a cautionary approach to the assessment of individual risk is appropriate or too many people will be excluded from a 'safe use' or informed choice.

Warnings, safety information, and instructions for use may be at a relatively high level of complexity, educational grade level or literacy, and understanding or comprehension. Readability testing (reading ease and understanding) often reveals that a significant proportion of consumers are simply excluded from effective communication as to appropriate use or hazard information. This is not universal design.

(h) Zero tolerance

Globalization, plus the Internet as a source of specific information, higher educational levels, and the availability of mass communication suggest that hazards, risk exposure, and injury reduction will become ever more important. Thus, what is currently acceptable in terms of personal injury may be drastically lowered. For some, universal design criteria may include very minimal or zero tolerated injuries.

The key concept of foreseeability should be achieved by objective means, followed by practical remedies that are tested to assure effectiveness. Mere compliance with current standards and regulations probably would not define an acceptable standard of care in the immediate future, particularly if a broader array of people expect or seek to avoid preventable injury and they are able to obtain technical product information. *Knowledge can be a powerful weapon, for both the consumer and the producer.*

Those who might object to universal design, with its more objective foreseeability and more specific standard of care, should remember that there is always a gradual transition to what, at first, seems to be the harshness of a new law or the burden of a new engineering design. Such transitions are generally marked by exceptions, variances, and extended compliance dates. The words used to soften the requirements or ease the burden include reasonableness, practicality, technical feasibility, cost viability, and the use of cost–benefit or other balance-of-interest phrases. However, there is a clear trend and objective that is in the overall best interests of society, so the question is how to achieve fully that which is now or will be required.

(i) Affirmative action

Universal design may be perceived, by some, as an added burden on the producer of goods. That is a short-term view, since it serves strong long-term goals. In terms of quality, there is greater customer satisfaction by more purchasers. In terms of safety, there are fewer injuries and liability claims. In terms of government, there is less need for regulation. In terms of society, there is a public relations benefit for the producer in being a good citizen. To achieve the goal, the following affirmative steps should be undertaken.

1. Get to know the real (actual) market of people and circumstances of use by direct contact and evaluation. Define the objective in terms of design requirements, company objectives, and legal obligations.
2. Design 'for all' includes testing for all relevant variables and attributes. Clearly define the limits and exclusions. Warn about the limits and inform about the

marketplace exclusions. Provide for adjustable features, conduct preliminary demonstrations and on-site testing, and include human error design compensation.

3. Maintain marketplace contact, through audits, service, or upgrades. This responsibility improves knowledge of the marketplace, helps to institute timely product improvement, and assists in recall, notices, and claims prevention.

4. Compliance with relevant trade standards and government regulations should not be relied upon as achieving a desirable standard of due care. Standards cannot cover all aspects of a design; they reflect a compromise to achieve consensus or trade acceptability, are only the first steps in helpful guidance, and do not reflect universal design principles. More flexible process requirements, such as techniques for risk assessment, can include universal design.

5. Understand that quality of life objectives require more benefit than detriment to the individual consumer who is included in the universal design envelope. Quality of life usually cannot be obtained when there are substantial unknowns in the standard of care or poorly communicated restrictions on use.

6. Anticipate that there will be increasing protections afforded to consumers, including legal liability, both as a form of redress and punishment and also to affect the market share or financial viability of an offending economic enterprise.

7. What is reasonable universal design varies as to the product, process, or service. 'Reasonable' includes foreseeability of those who might be harmed, in performance or injury, with an appropriate standard of care clearly established for each product or system.

(j) Costs of universal design

As to the cost of universal design, it depends on whether the process is cut-and-paste, reactive, or proactive. The most costly is *cut-and-paste*, which is a series of design, test, redesign, and retest events. As one chief engineer explained, when there was a major failure just before the production schedule was to begin, production had to be given priority. Remember the old adage that 'production is king and it pays our salaries'. Another project engineer indicated that the learning process had begun as production started, and that they would focus on corrective action for test failures and field complaints as they occurred. It is a reality that market pressures may force an early release of unfinished or unpolished designs. A *reactive approach*, particularly to design safety, involves responses to major accidents and serious claims, by use of remedial retrofits, recalls, notices, instructional bulletins, and an active liability defense posture. It may involve on-site safety assessments which gradually evolve into a universal design, but it can be costly. The most economical approach is the *proactive approach* in which original design requirements are carefully formulated (EEC 1989, COM 1990, ITIIC 1994), risks identified and fully assessed (EEC 1989, COM 1990, HS (G) 48), lessons learned integrated with objective information, and a universal design achieved early in the process. Quite simply, this is a form of 'preventive medicine' that saves money.

(k) Contractual compliance

If the statement 'universal design principles apply' appears in a procurement or purchasing agreement, the question may arise as to what proof there is of substantial compliance with the contract. By definition, the design is universal, so it should accommodate a very broad population. Even if specifically restricted to an identified specialized group, there still may be attributes and characteristic variables that are foreseeable and relevant. How are they identified and prioritized? What recorded analyses and tests are reasonably necessary? In other words, years later is there something in writing that will show the exercise of reasonable care under the circumstances and prove contract compliance?

The effort to achieve universal design is multidisciplinary, including system safety (fault tree, failure mode, and similar analyses) and human factors (such as human error prevention and control of residual risk). Since research is also necessary, how is this best integrated into the organizational structure of the company? Do the decisions in design reviews have an appropriate foundation in objective facts, based on various forms of research and investigation? Is it only a marketing perspective as to the product purchaser, a limited view that ignores resale and recycling, or a focus on component rather than system effectiveness? What criteria have been used for hazard identification and risk assessment (Peters and Peters, 1999a)? Remember, tobacco cigarettes are legal, have had a rather high level of acceptable risk, and have liabilities attached to their sale around the world. For each product or system, there may be unique considerations as to safety, health, and the environment. Is there a well-considered overall policy and plan from which appropriate design tasks can be derived to achieve a universal design acceptable to others? The basic caveat is: how will it be possible to demonstrate that there was a reasonable standard of care for a more broadly defined user population?

(l) Simple products

A review of what might be needed to achieve a meaningful universal design may seem to impose too great a burden, particularly on small enterprises producing simple products. However, simple products may be small systems, may be incorporated into large systems, may become a vital link in overall performance, or may result from a delegated design function from a major assembler with inherent obligations. Universal design may be repugnant to some in terms of risk-taking choices, competitive cost-cutting decisions, or the exercise of personal freedom in the production of goods and services. However, design for safety is predicated on and derived from customer specifications, hazard analyses, government regulations, trade standards, and the common law applicable to all. It is not wise to rely on apparent distinctions between the legal obligations of small, medium, or large organizations involved in commercial, industrial, military, or government projects. Such distinctions are often too minor, over-sophisticated, somewhat transitory, or applicable only to one cost–benefit situation. Thus, the burden is on all companies and entities to determine exactly how they might meet the design challenge in a cost-effective manner. It is also a challenge for the individual professionals who may have to devise and undertake plans to accomplish the objectives now being thrust upon them in the name of universal design.

(m) Conclusion

There are increasingly complex technological changes which now require greater customer reliance on the supplier who places goods in a marketplace, particularly in an era of world trade and diverse populations. The consumer or user may believe that there is some legal right, moral entitlement, or ethical expectation for a reasonable freedom from harm relative to safety, health, and the environment. This assumes that there has been an application of some engineering standard of care, such as universal design, for the equal benefit of all who might come in contact with the product, unless specific exclusions or restrictions have been communicated to those who might be adversely effected. Equal protection of the law suggests that the engineering standard of care should be derived from, or made compatible with, all applicable legal requirements. In an age of world trade, some harmonization of the applicable laws of producer countries would seem desirable for either equal protection under the law or equal benefit from engineering design considerations. One first step in meeting societal and purchaser expectations is the acceptance and promotion among engineers of the concept of universal design as a reasonable standard of care.

Caveat: The automotive industry deserves great credit for the vast number of vehicles being operated by so many people in so many countries, for so many practical and useful purposes. Some automobile enthusiasts do not like any user-friendly concepts, because such concepts seem to suggest a cetain blandness, a fostering of design objectives that might take the fun out of driving, efforts that could result in removal of their high performance and muscle cars, and a narrowing of individual choices at time of purchase. Some may take pleasure in the opportunity to demonstrate their unique and superior driving skills by maneuvering among other vehicles on fast traffic highways, so they want to preserve their passion for the thrills of risk-taking. Some marketing specialists want to promote and advertise the adventure and pride involved in driving a particular brand, so they fear what seems to them a homogenization or harmonization that could blur marketing differences. These may be specious arguments, but are heard and have been persuasive in the industry. Therefore, universal design is a concept that grows rather slowly in popularity, and for each further incremental advance it requires considerable skillful research and innovative creative engineering design.

7 The distracted driver

(a) Introduction

A serious heath problem is developing from automobile collisions caused by distracted drivers. This is the result of the rapid proliferation of portable cellular telephones and personal organizers used while driving (inattention to road traffic conditions), the development of more sophisticated entertainment systems and instrument panel controls (less vigilance), the advent of navigation and television displays in vehicles (eyes off the road), and promises of sophisticated wireless e-mail, fax, and Internet services in the vehicle (addition of complex activities). Preoccupation with electronic gadgets may degrade human driving performance. But many drivers sincerely believe they have the talent to do several things at the same time, such as hold and look at a cellular telephone in one hand, drive with a beverage container in the other hand, and exercise their personal skills. Obviously, they feel that they do not need two hands on the steering wheel, two eyes on the road, and their attention focused on the traffic around them.

This is a unique situation requiring intensive health promotion because distracted or 'offensive driving' may be habit-forming and difficult to change, any significant design remedies will be slow to arrive and may be circumvented, and the regulatory laws have proved difficult or impossible to enforce. This special need may require research to determine the most effective techniques for health promotion.

(b) Distractors and risk reduction

(1) Distractors

An automobile driver may dramatically increase the risk of an accident if he is unaware of the special hazards involved in the use of certain vehicle accessory devices and aftermarket equipment. This includes the use of cellular telephones that could result in inattentive driving on crowded roadways; the increasing use of navigation displays on the instrument panel that could result in drivers looking at them rather than the ever-changing roadway traffic; the presence of captivating entertainment and interactive information systems including on-board television, fax, Internet, computers, wireless data and message communication devices, and other complex electronic devices that may tend to distract the driver from a focus on safe driving.

The key words are 'distractors' (devices) that create *cumulative* 'distraction' from the driving tasks that can result in injury risks that are largely unrecognized and unexpected by the driver. The safety problem may become worse if proper research efforts, to minimize risks, are not conducted by the manufacturer and the information on residual risks is not effectively communicated to both the dealer and the vehicle driver or owner. In general, health promotion has become a critical factor for the successful introduction of advanced technology for automotive vehicles.

(2) Risk reduction

For electronically 'wired cars', some marketers and dealers insist that there can be no substantial distraction because the driver is very capable of 'multitasking' (dealing with several sources of information simultaneously) while the vehicle is in motion. They may believe they can demonstrate this ability in a moving vehicle. Some human factors specialists insist that 'cognitive tunneling' can occur and the driver could momentarily lose track of close-in traffic (for safe navigation) while concentrating on a cell phone call or other electronic device (a distractor of attention). Such distinctions might be important to research and in accident reconstruction, particularly if the owner or dealer has added some aftermarket in-car electronic devices that could act as primary, secondary, or cumulative distractors.

There are three obvious risk-reduction techniques for this problem situation. *First*, there could be printed and audible warnings not to use a device unless the vehicle is stopped. For example, a front center-mounted computer may have a written warning, in the operating instructions, to use it only when the vehicle is stopped and in a parking space, lot, or area. However, if the computer can be viewed while the car is in motion, some drivers will predictably do so. Similarly, a recorded voice warning to use a fax or to display an e-mail only when stopped may not be thought applicable to passengers. It is not uncommon for the driver to be distracted by or become involved in the passenger activity, thus becoming degraded in terms of the primary driving tasks. The *second* technique is to have an integrated system, with all devices either cut-off from or out-of-sight of the driver when the vehicle is in motion except for warnings, heads-up navigation prompts, or customary radio signals. This design approach might categorize such possible distractors as primary or secondary in importance, with needed information highlighted for quick cognitive access and understanding, and with warning overrides for commandeering attention. An adaptive cruise control might, at first, provide advisory information, then a priority warning, and finally could automatically slow or steer the vehicle if an object detected by radar or other sensors could cause harm. The *third* technique is marketing and dealer restraint. Dealers should recognize that prospective vehicle purchasers vary greatly in intelligence, comprehension, attention, perception, encoding and resolution of competing information, and executive-type decision-making. This serves to illustrate that risk reduction should become more of a priority for marketers, dealers and others in the future. This may be helped by increased dealer consolidation, restructuring, electronic communication, and technical sophistication.

The new responsibilities of the automobile marketer and dealer should be a health promotion concern. There is much that only the dealer can accomplish, but knowledge and motivation must be fostered by public promotion efforts.

The consumer is well aware of the surprisingly low price for multi-function wristwatches, calculators, and other consumer electronics devices. Cost of the microelectronic devices should not be a major restraining consideration for mass-produced items.

(Frank, 1993, p. 267, speaking about vehicle external interactive highway guidance, control, warning, and signal devices and systems for automobiles)

(3) Magnitude of the risk

In recent years, there have been many statements (opinions) that the use of wireless telephones in moving vehicles has caused accidents. The Japan National Police Agency (JNPA, 1996) asked drivers not to talk on mobile telephones while driving. Other countries have attempted to limit the use to hands-free telephones because they believed there was some danger.[1] Many local government agencies have attempted to institute bans or prohibitions (such as use only when the vehicle is stopped or during emergencies). However, in the United States a federal agency, in 1997, concluded that insufficient data existed to estimate the magnitude of the problem, so no action could be taken (www.nhtsa.dot.gov/people/injury/research/wireless). This may seem to illustrate the fact that there are differences in opinion as to whether there is a danger necessitating corrective action. But the government policy in the United States is that there must be convincing field data, even if reliable data might be difficult to obtain, rather than applying the precautionary principle utilized in the European Union, where a possible danger raises a 'red flag' (Foster *et al.*, 2000; Raffensberger and Tickner, 1999) and risk assessments are conducted differently (COM, 1990). In essence, the hazards (propensity to cause harm) may be recognized, but the actual danger (magnitude of the risk) may be disputed.

In July 2000, in the United States, there were radio and television reports with information attributed to the American Automobile Association indicating that millions of collisions a year could be blamed on the distracted driver. There were similar opinions, from others, that more than 25% of all accidents were caused by distracted drivers. One report indicated that a driver using a cell phone was believed to increase the risk of an accident by a factor of 4 and, if he was drinking, it would increase the risk by a further factor of 4. These were primarily statements concerning a probable hazard, and opinions as to possible risk, but they did not advocate or express a need for a remedy.

Press releases stated that vehicle collision warning systems were being tested, with a radar range of 500 feet, that would provide about 0.5 to 1.0 second of additional time to avoid an accident. It was estimated that this device would result in a 50% reduction in some types of accidents. This suggests the importance of one second in driver reaction time.

---

1 The use of hand-held cell phones in motor vehicles has been restricted in Austria, Brazil, Chile, Denmark, Germany, Greece, Israel, Italy, Japan, Poland, Portugal, the Philippines, Romania, Singapore, Slovenia, South Africa, Switzerland, Turkey, the United Kingdom, and in the state of New York. But there are questions about enforcement, the habit and convenience of such use, and what constitutes an emergency exception to the rule forbidding driving with a cell phone next to the ear. Is distraction reduced with headsets, voice dialling, the use of the vehicle's radio speakers, and other hands-free accessories?

The safety objective may be eyes-on-the-road, hands-on-the-steering-wheel, full attention on the driving task, and an appropriate level of information processing by the driver. These issues are undergoing research by government agencies, universities, and by the companies that are introducing more and more complex electronic devices into automotive vehicles.

Dangerous events can materialize very quickly, as indicated in Table 7.1. For example, a vehicle can make a quick lane change (in 2 to 3 seconds) and slow down in front of a distracted driver (with a blind time of 4 seconds) and a collision might occur. Similarly, a distracted driver might be following too closely (with insufficient headroom) when the vehicle in front initiates a sudden braking action to slow or stop the vehicle. Perhaps more serious is the business person who is using the vehicle as a mobile office on wheels and becomes distracted by attempting to interpret a garbled or fading voice communication or attempting stock trading or appraisal. There could be a complete loss of overall situational awareness and virtually no readiness to respond to some of the situations shown in Table 7.2. Incidentally, Table 7.2 relates to behavioral constructs, whereas Table 7.1 is based on recent findings in biochemical neuroscience.

Many drivers have their heads oriented so they are keeping their eyes on their roadway path of travel, yet are actually driving while distracted (DWD). They may believe that they can react normally should a danger arise. Under such circumstances, the issue is whether there may be tunnel vision (restricted peripheral vision), limited search (of the field of view), degradation of visual quality (impaired detection), or a total loss of situational awareness (a temporary ignorance of what is happening around the vehicle). The driver's belief may be misleading and be an expression of controlled risk-taking behavior, but can foster attitudes antagonistic to a true understanding of the limitations of human capability involved in that situation.

During the past 30 years there have been many studies of mobile telephone use in automobile vehicles (e.g. Alm and Nilsson, 1994; Briem and Hedman, 1995; Brookhuis

Table 7.1 Human reaction times (highly variable) (seconds)

Activity	Situation		Commonly utilized	Source
Perception (detection and awareness)	Simple Complex	0.5 3.0–4.0	1.5	AASHTO (1973)
Reaction (braking)	Simple Complex	0.5 1.0	1.0	AASHTO (1973)
Swerve (avoidance)		0.9–2.0	1.5	Johansson and Rumar (1971) Hulbert (1984)
Maneuver (passing)		3.5–4.5	4.5	AASHTO (1973)
Preview (scene)	Look ahead Look back	2.0–2.5 0.8–1.0	2.5	Hulbert (1984) Robinson *et al.* (1972) AASHTO (1973)
Headway (distance)	60 mph (96 km/h)	1.0	1.0	Hulbert (1984, 1976) Robinson *et al.* (1972)
Search (visual)	Lane change Enter crossroad	0.8–1.6 1.1–2.6	0.8 2.5	Robinson *et al.* (1972) Hulbert (1984)
Sight distance (hazard detection up to braking)	Legal assumption 95th percentile	1.6	0.75 2.5	Hulbert (1984) Olson and Sivak (1986)

Speeds for pedestrians: actual 2.5–6.0 ft/sec (AASHTO, 1973); design 3.0–4.0 ft/sec (*Manual on Uniform Traffic Control Devices*, 1988). Side wind gust correction. Conversions: 1 mph = 1.467 ft/sec = 1.609 km/h = 26.8 m/min.

et al., 1991; Brown *et al.*, 1969; Fairclough *et al.*, 1991; Kames, 1978; McKnight and McKnight, 1991, 1993; Maclure and Mittleman, 1997; Pachiandi *et al.*, 1994; Parkes, 1991, 1993; Petica and Bluet, 1989; Petica, 1993; Redelmeier and Tibshirani, 1997; Spelke *et al.*, 1976; Stein *et al.*, 1987; Sussman *et al.*, 1985; Violanti and Marshall, 1996; Wang *et al.*, 1996; Zwahlen *et al.*, 1988). Despite such illustrative publications, there is still much to learn, perhaps because of the differences between what drivers 'could do' in laboratory and simulation tests and what they 'actually do' in on-road testing and typical driving situations. One benefit from distractors may cancel another. For example where tedium and fatigue could occur in continuous and monotonous truck driving, a secondary voice communication task was found beneficial in terms of performance (immediate alertness and concentration), but had a fatiguing effect on the driver (Drory, 1985). It has been known for some time that multiple tasks can result in performance degradation (Noble and Sanders, 1980).

When the vehicle driver is a single-channel information processor, the driver's attention may be rapidly shifted between tracking and object avoidance (Hulbert, 1984). For the tracking task on a committed pathway, the driver 'looks ahead' (preview time) for about 2.5 seconds (AASHTO, 1973). For a safe lane change, the 'look back', by head turn or rear-view mirrors, is about a second or more. In essence, the time allowances for forward vision and for preparation for lane change are relatively small compared with common distraction times. That is, during distraction time (blind time) objects can suddenly appear and become collision-prone.

The reaction time advantage of a heads-up (windshield) display may be appreciable when compared to an instrument panel display. The eye travel time may be half of that of a look-down movement (head and eyes) combined with a simple visual search, perception, and comprehension task. However, there is considerable variability among drivers, and even slight complexities in the visual task greatly escalate the time requirements and safety implications. For example, the difference between 2, 4, and 8 seconds (blinks of an eye) of off-the-road vision can be translated into 176, 352, and 704 feet at a 60 mph (96 km/h) vehicle speed. Assume a crowded highway and what another vehicle might do, and those hundreds of feet in a few seconds with eyes off the road become important. Even on a rural or country road, with vehicles traveling only 30 mph (48 km/h), what about an intersection with a vehicle approaching on a side road? For 8 seconds, your eyes are off the road for 352 feet of travel while the other vehicle also travels 352 feet toward the intersection and it may come into your view and pathway during the blind time. In other words, logic suggests that just a few seconds may appreciably increase the risks of accidents. Thus, the key design requirement is to minimize off-the-road distraction time by appropriate display location, cues and alerts, search time, perceptual simplicity and accuracy, and relevant comprehension.

The health promotion efforts should include demonstrating the blind time hazard (eyes off the road) and just how easily time extension (added seconds) can occur. The driver can be advised to practice visually locating displays, how to make quick inter-pretations when necessary, and when to ignore purely entertainment features. Advanced cell phones (WAP), with access to the Internet, e-mail, online games, and other services should not be used by the driver when the vehicle is on the roadway.

The use of cell phones varies greatly as to roadway location, the percentage of the local population owning cell phones, and local laws regulating their use while driving. In one urban location, some 8% were driving while distracted. The *individual* increase in relative risk of 4 would result in an *overall* relative risk of only 0.32. This suggests that cell phone usage, while driving, may be acceptable, but this does not include the cumulative and interacting effects of other distractors that are just beginning to appear in new vehicles and are being offered for installation on older vehicles. Cumulative risk may be added, but interactive risks may be multiplicative. Research is needed, but health promotion may be quicker because of the diverse mix of vehicles, current distractors, and future advanced technology distractors. The driver should focus on the road, not on the conspicuous use of expensive electronic devices.

(c) Information processing

The human information processing system can deal with vast quantities of data, such as the ever-changing flow of visual information presented to the driver of a fast-moving vehicle on a crowded scenic highway. Obviously, humans have to have a built-in means of preventing an information overload and a process to select that which may need prompt attention (see Table 7.2). There are brain mechanisms that filter out non-relevant clutter, suppress multiple competing stimuli, and direct attention for fast reaction (Kastner *et al.*, 1998; Henson *et al.*, 2000; Johnson-Laird *et al.*, 2000). The design of a vehicle, including all accessories and options, should not challenge or be antagonistic to the biological mechanisms of human response and survival. In essence, it should not significantly degrade reaction times, interfere with identification accuracy, or disrupt essential cognitive functions. It should enhance directed attention when necessary for injury avoidance, assure the accuracy of perceptions and interpretations, help facilitate priming, coding, and executive functions, and assist in minimizing undesirable distractions, conflicts, and confusions. It should be apparent that this design-oriented approach requires considerable interdisciplinary research, engineering development, and risk assessments that require substantial time and effort. Thus, there should be some other, more immediate, action, because the accident injury risks are already high, are rapidly increasing, and could become virtually unmanageable in the near future with the proliferation of technology-driven distractors.

The distraction time (eyes off the road) is variable according to the task performed. It should be minimized and controlled for safe driving. However, the current critical questions relate directly to information-processing capabilities. *First*, under what circumstances might there be separate, shifting, or shared attention to the driving task during distraction? Is there time-batched shifting, proportional sharing, or a delayed

Table 7.2 Distraction stages

Stage	Phase	Activity	Brain function
1	Preliminary (arousal)	Expectations (goal established)	Code selection (local, sparse, and bias for neural circuits)
2	Conduct (directed attention)	Search (locate and fixate)	Visuo-motor-spatial (guidance)
3	Perception (understanding)	Identification and readout	Filtering, code matching, and fact acquisition
4	Comprehension (meaning)	Interpretation in context	Associative interaction and network recruitment
5	Verification (achievement)	Subsequent choices	Executive judgment

Time required: primed ~2 sec, novel ~4 sec, complex >8 sec.

single focus? In other words, are there moments of blackout and blind driving or periods of degraded driving that might or might not be tolerable under the circumstances? Stated another way, when one is distracted by a primary task, is there any meaningful awareness or attention relating to the secondary driving task? Could there be a mental model of the secondary scene where unexpected changes would alert a driver? If so, how timely and effective would such secondary or peripheral perception be for simple and complex tasks? *Secondly*, what are the effects of the driver's head position, the distractor location, the distractor target complexity, practice and priming, and hazard conspicuousness? Appropriate research might reveal how distractors could be made more acceptable and testing might provide a risk rating for each distracter.

> Cellular telephone activity was associated with quadrupling of the risk of a motor vehicle collision . . . we observed no safety advantage to hands-free as compared to hand-held telephones.
>
> (Redelmeier and Tibshirani, 1997)

> Results indicate slower brake response times to the change in the traffic control device (onset of a red light) in the presence of the in-vehicle distractor. However, drivers exhibited significantly shorter stopping times to this red light activation in the presence of the distractor. Despite this latter effect, the margin of safety, as represented by stationary distance from the intersection, was significantly reduced approximately 25% in the presence of the distractor.
>
> (Hancock *et al.*, 1999)

(d) Health promotion

Unlike many convenience and entertainment features that are built into a vehicle, the modern cellular telephone is portable and can be acquired independent of the vehicle. Anyone can bring a mobile cellular phone into an automobile and external control of its use would be rather difficult. When there were few cellular telephones and their cost of operation was high, the overall risks were low. Now, an increasingly high proportion of the general population has such telephones, they are wireless, the operating costs are low, and they satisfy an evolving communications dependency for both business and personal needs.

The latest generation of such devices provides Internet services that demand greater attention and reading time. The plasma and liquid crystal displays are becoming larger and more capable of high-resolution picture transmission and display. They also serve as personal organizers, so calendars and schedules can be scrolled into view. There are attempts to make them attractive miniature personal computers. In other words, the distraction problem does have future ramifications and probable risk increases.

Legal attempts to regulate the use of distractors while driving are much more difficult than for the more conspicuous and common speeding violations. The cell phone also has emergency, medical, location, and security benefits. But what will constitute an emergency or security problem in the opinion of the driver who believes cellular phone operation while driving is a necessity?

In addition to attempts to regulate distractor use, there have been attempts to design-out much of the risk by some major manufacturers or, at least, to develop some consensus design standards. At the same time, the small aftermarket designers have done the opposite by installing devices, such as television monitors, that can be viewed by both the driver and passengers on long trips. Thus, design improvement and legal restrictions may have limited and long-term effect. The only immediate approach to significant risk reduction is an effective health promotion effort by many sources in many repeated forms (Peters and Peters, 1999a). Special care is needed or short messages may have an opposite or unintended effect, particularly among the young and rebellious or subpopulations prone to overt risk-taking conduct (Peters, 1991). Tobacco (cigarette) marketing illustrates the difficulty of formulating an appropriate and effective heath promotion effort. Merely conveying the fact that distractors result in accidents will not appreciably reduce the risks when there are perceived benefits or incentives to use the distractors.

It can be argued that consumers wanted beverage container holding devices (cup-holders) in automobiles and this led to one-arm driving while drinking the beverage. Further, the beverage led to eating sandwiches and other solid edibles. Similarly, visor mirrors increased driver primping, grooming, and looking into the mirror while driving. Conversely, on-board radios have slowly evolved into manual push-button, station self-searching, and quick-mute devices that do not seem to be appreciable distractors when used properly. The basic problem is that drivers have become accustomed to the growth of various kinds of distractors and want to utilize the latest devices, which could have an adverse cumulative effect on distraction. Thus, health promotion faces formidable obstacles in the general acceptance of the distractor situation.

Since the need for 'education' is so important, shouldn't some intensive and varied research projects be undertaken to discover what might be the best (most effective) methods to promote health in this unique situation? Remember, the danger is not restricted to the driver who undertakes undesirable behavior, since the driver's conduct places many other vehicle drivers and pedestrians at risk. It is an unnecessary and avoidable risk that could be reduced significantly by appropriate health promotion that is timely undertaken. If large financial resources are being applied to develop devices that distract, why not apply a rather small fraction of that amount for independent health promotion?

(e) Future research

Research on driver distraction has been conducted in simulators, on test tracks, by laboratory experimentation, by observations of on-road practices, and by accident data analysis. The emphasis has been to gain information to help develop products that can reduce the potential for accident-inducing distracted driving. Far less has been done on the research necessary to fully understand the cognitive or brain processes that must be accommodated or developed for the products (see Table 7.3). There are now many imaging systems that can reveal patterns of gross brain activity in response to distractions in the mental workload. At a more precise level, the complexity of the brain processes can be inferred from the fact that the human brain contains over 100 billion nerve cells, each directly communicating with 100 to 1000 other target cells, and using any of 100 different neurotransmitters. The synaptic responses may be fast acting (under one millisecond) or slow acting (hundreds of milliseconds), excitatory or inhibitory, and short-term (synaptic specific enhancement) or long-term (cell-wide changes). Much understanding has been gained by modern neuroscience and even by molecular biological examination of the activities within a particular nerve cell.

It is known that much of driver performance is based on the *implicit memory* where perceptual and motor skills are expressed without reference to conscious recall, and this

Table 7.3 Distraction effects (for each distractor in play)

Mental status	Driver perceptions	Attention to driving task	Neural basis	Effect	Fault
Passive	Unaware of any distractor	Focused	Facilitated code associations (matching)	Efficient fact acquisition, expectancies, and choice behavior	None discernible
Passive	Not distracted	Attentive	Efficient circuit functions	Enhanced function	None discernible
Active	Aware of distractor in background	Attenuated focus	Variable suppression (ambiguity)	Some degradation	A contributing cause
Active	Partially distracted	Shared attention	Excess circuit functions	Attenuated functions	A contributing cause
Responsive	Interactive behaviorally or emotionally	Multitasking; possible overload or oversight	Channel switching (conflicts)	Filtered situation awareness	A substantial factor
Responsive	Substantially distracted	Attention diverted	Inefficient brain functions	Confusion	A substantial factor

Table 7.4 Learning stages

Stage	Activity	Speed	Location
1	Synaptic neurotransmitters	Immediate	Intercell
2	Physical plasticity	Rapid	Cell surface
3	Vesicle recruitment*	Fast	Intracell
4	Circuit enhancement	Gradual	System
5	Morphogenic changes	Slow	Cell

In terms of cognitive and brain research, there are learning stages involved in the use of complicated, multifunction, densely integrated displays and control systems. These learning stages involve physical and chemical changes in the central nervous system.

*Axon and dendrite.

happens very quickly. The *explicit memory* involves conscious recall of past objects, places, and events (that is, the hippocampus function). These memories are affected by *sensitization*, so some responses occur to rather neutral stimuli. One ultimate research question is whether or not distractors affect sensitization, explicit memory, or implicit memory and their associated learning processes. There will be many other questions relating to human response characteristics under various driver distraction modes. Hopefully, the research will help to solve the current unknowns in driver distraction.

(f) Conclusions

Distracted drivers represent a significant public health problem that is resulting in an intolerable magnitude of personal injuries. It only takes one distracted driver to endanger other vehicle drivers, passengers, pedestrians, and bystanders. Correction and preventive treatment should be interdisciplinary and multimodal, with a key component being targeted health promotion.

Research is needed relative to the identification of road hazards. What is the perceptual capture efficiency under conditions of expectation (conformance to a mental model), the unexpected but predictable event, and in terms of sudden surprise? What are the effects of complexity and ambiguity? How could these factors affect roadway design, traffic regulation enforcement, and driver training?

Health promotion research would be helpful if it related the driver's awareness of the blind-time hazard (eyes off the road), recognition of time extension (excessive lost time in distracted concentration), and understanding the degraded perception (from rapid shifting of attention). The objective would be to determine how the driver could be properly informed or adequately educated as to the meaning of specific facts and the need for social conformity while on the public roadways. The basic objective of health promotion would be to determine the most effective, economical, and long-lasting techniques for changing driver attitudes and behavior in relation to to safe driving and crash injury reduction.

> We believe the cellular-telephone industry now has an ethical obligation to include warnings and advice with their products and mailed bills; to support and assist in further engineering, ergonomic, and epidemiologic research; and to provide easy-to-dial toll-free numbers for reporting road hazards and unsafe driving. The last step would help cellular-telephone users contribute to roadway safety and compensate for the additional hazard incurred by cellular-telephone use.
>
> (Maclure and Mittleman, 1997)

8 Occupant injury protection: biokinetics

(a) Introduction

Modern technology has produced automotive vehicles that have become both a luxury and a necessity in modern civilization. They have become highly useful, ever more varied in form and function, and capable of high speeds on crowded roadways. One unfortunate consequence is the high frequency of accidents and the greater severity of injuries when collisions do occur. In response, modern technology has produced a variety of safety and health features, devices, and designs intended for better occupant protection in high-speed vehicles. Injury reduction has become a prime design objective, but there are residual risks, which, as technology evolves, require effective communication to those at risk. There can be little risk-avoidance behavior without awareness of the hazards and effective communication to the vehicle occupant as to what could and should be done for self-protection. For example, one out of three drivers apparently fails to understand the function of head restraints, few understand the 'safe zone' posture required for airbags, and many believe safety features should be adjusted only for comfort. Some of the current residual injury-producing problems in occupant systems are specifically described here in order to illustrate what is needed in terms of both design remedies and health promotion activities.

The magnitude of personal injuries from automobile collisions is well known to trauma specialists, treating physicians, legal specialists, insurance claims personnel, and the general public. With increasing vehicle speeds, the number and variety of vehicles, and the diversity of drivers, there is an urgent need for additional injury reduction actions. The automobile industry as a whole has been attempting *design remedies* such as improvements in crashworthiness (energy management) and occupant protection (e.g. seatbelts, airbags, and compliant interiors). Unfortunately, far too many vehicle drivers do not understand the reasons for vehicle safety features and, as a result, fail to take the necessary action to assure their own safety. Vehicle occupants, through lack of knowledge, may lower the health and safety benefits or defeat the purpose of safety devices, or create new hazards and injury vectors. In terms of *health promotion*, there should be better communication of the specific purposes of occupant protection features and what deliberate actions are needed on the part of vehicle occupants to assure injury minimization in what has become a reasonably probable chance of an accident scenario for most drivers in their lifetime.

(b) Proper use of head restraints

During an automobile collision, the driver's head may be subject to violent forces from different directions. The vehicle itself may come to a sudden stop in a frontal collision, be pushed to a higher speed from a rear collision, exhibit complex motions from rollovers or a side impact at different locations on the vehicle body, or strike another object on the roadway. The human body continues to move at its original velocity and in its original direction until changed by interior contact. The seatback serves to protect and cocoon the upper torso in rear collisions. The seat belt restraints limit motion in a frontal collision. However, the head may bounce around in an unrestrained fashion since the head mass movement is dampened only by relatively weak neck muscles and ligaments, neural and vascular tissue, a flexible vertebral spinal column, and bone-to-bone contact. The injuries occur from biomechanical forces categorized as causing hyperextension, hyperflexion, axial rotation, lateral bending, compression, tension, decelerative forces, blunt trauma impact, or a complex combination of several such forces. The injuries may be varied, but could include vertebral dislocation, spinal cord injury, nerve root and ganglia damage, hematoma and necrosis, paralysis, various forms of brain injury, and may result in death.

The primary objectives for injury reduction are: (1) to keep the head and upper torso in alignment (prevent 'head lag'), (2) to reduce peak accelerations (by 'energy management'), and (3) to minimize unusual interior contacts (point loading and secondary movements). Alignment is achieved by appropriate head restraints and airbags (Figures 8.1–8.3). Acceleration is controlled by stress distribution and absorption. Interior contacts are reduced by self-centering, seat contouring and wings, and soft broad interior fixtures.

Expanding upon a recent study of head restraint height (Peters, 1999), to include drivers in Canada (Table 8.1), it was found that about one out of three drivers would

Figure 8.1 During a rear collision, a low head restraint may permit the driver's head to rotate abruptly rearward, resulting in hyperextension injuries to the neck and deceleration injuries to the brain.

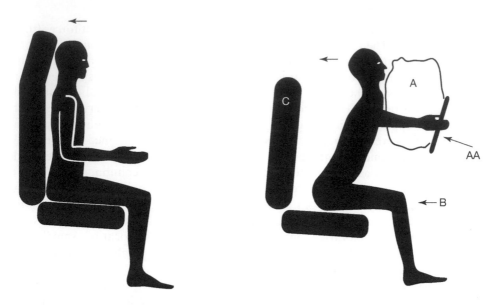

Figure 8.2 (Left) A high head restraint reduces violent head motions and keeps the head and upper torso in alignment.

Figure 8.3 (Right) In a frontal collision, an airbag (A) should cushion the relative forward motion of the head and upper torso, together with the effect of a crushable steering wheel (AA). Upper torso rebound from the airbag and lower torso movement from the action of the knee bolster (B) should return the driver to an erect position on the seatback and head restraint.

Table 8.1 Findings of head restraint study

Category/height	USA (Ohio)	USA (California)	Canada (Ontario)
Safe[1]/high	18%	18%	14%
Marginal[2]/medium	46%	51%	52%
Unsafe[3]/low	36%	31%	34%
Sample size (N)	162	306	443 [4]

This study did not encompass the distance of head from headrest (which should be approximately 1 inch), deal with excessively narrow restraints, deal with head restraint configuration deficiencies, catalog out-of-position occupants, or account for adjustable headrests that had been removed.

[1] Safe is defined as a height near the top of the driver's head.
[2] Marginal is defined as a height near the ears of the driver.
[3] Unsafe is defined as a height near or below the shoulder level (where there is unfettered head rotation in a collision).
[4] 6% of the drivers in Canada had a head restraint height more than one inch below the shoulder line.

have had no restraint of head movement in the event of a collision into the rear of their vehicle, thus becoming prime candidates for hyperextension neck injuries. Fewer than one in five had adequate head restraint.

This was based on the observation of 911 vehicles on high-speed roadways, driver position only, in 1999. 'Safe' was defined as a height near the top of the driver's head. 'Marginal' was defined as a height near the ears of the driver. 'Unsafe' was defined as a height near or below the shoulder level (that is, a totally unrestrained head rotation in a rear collision).

An inquiry of drivers revealed a belief that the 'head*rest*' was only a *comfort* feature. There were, however, some vague impressions that the head restraint might have something to do with safety. Few understood how to adjust an adjustable head restraint to a safe position. None knew that the head should be positioned about 1 inch (2.54 cm) from the head restraint. None had considered centering the head on the head restraint (that is, avoidance of an out-of-position or off-center location). Those who had a fixed position head restraint (integrated into the seat back), whether high or low, wide or narrow, well configured or not, believed that what the manufacturer had supplied was safe for them. In a study of parked vehicles, a few vehicles were found where the head restraints had been removed. When drivers were questioned as to warnings and instructions in the owner's manuals, there was an almost complete lack of awareness, communication, or understanding that could have alerted the driver to the need for proper head restraint.

A failure of communication suggests that remedial action should be taken by government entities, trade associations, and companies as a public service. A short videotape could be shown to new customers by the dealer and could also be sent to prior vehicle owners. Owner's manuals should be more explicit and specifically identify the hazards and risks of a particular vehicle's occupant protection system. The purposes should be to explain why the safety features are necessary, how to make adjustments, why body and head position is so important, and to enable appropriate risk avoidance behavior by the vehicle owner (Peters, 1999). The vehicle purchaser should have enough information, with the help of the dealer, to be able to determine whether a particular vehicle is 'safe' after all seat adjustments have been made (since weight, leg length, torso height, seat configuration, pedal height, intentionally out-of-position situations, and other variables may all have a surprising effect on head position in a particular vehicle). This is the 'informed purchaser' principle based on assisted learning and decision-making.

From a design standpoint, it is preferable to have a fixed (non-adjustable) head restraint in order to avoid the 'human error' of improperly adjusted head restraints. There are also dynamic or articulated head restraint devices, which move upward and forward in a rear collision, actuated by the forces generated by the upper torso or pelvis (that is, an automatic device deployed at a time of need). The seatback contour and wings should self-center the occupant and the head restraint geometry should self-center the head.

(c) Airbags

Another topic in which there are significant communication problems that need to be remedied is the correct use of airbags for occupant protection. For example, many vehicles have a tilting steering wheel for three reasons: (1) to tilt up for additional room to exit or enter the vehicle, (2) to tilt down for drivers of below-average height, and (3) for continued readjustments on long trips to relieve tensions and fatigue. Usually, the written instructions are to tilt the steering wheel for the most comfortable position for all driving conditions. Unfortunately, the steering wheel also serves to aim the airbags. It should point to the chin to decelerate both the head and chest (upper torso) at the same time, but a small misdirected bag can deliver its force to either the chest or the head separately. The new larger airbags, with a greater pillow or cushion effect, must have a greater inflation rate (to achieve the greater volume, while maintaining the required timing, in a limited space), thus may move faster and with more force being directed at the occupant.

The timing of the airbag inflation is critical, usually about a 15 to 20 millisecond crash sensor detection time and another 30 milliseconds to inflate fully, so that the occupant strikes the bag when it is fully inflated and starting to deflate (vent). This allows the head and shoulders of a 50 percentile male, in a vehicle traveling at 30 mph (48 km/h), to travel 5 inches forward to hit the air bag at about 14 mph (22.4 km/h) (with a rebound speed of about 5 mph) (8 km/h). In order to allow 5 inches (12.7 cm) of movement for both the driver and the airbag, the driver should be at least 10 inches (25.4 cm) from the air bag before it inflates. If the distance is too close, the inflating airbag will have an 'early hit' and slap, punch, or push the occupant rearward. This could dramatically increase the occupant's head and/or upper torso rebound velocity. Should the distance be too great, however, there can be a 'late hit' on a deflating air bag that is too soft to prevent contact with the hard portions of the steering wheel. The greater the early hit, either by a 'too close' occupant or a delayed or tardy airbag deployment, the greater the time for the occupant to be accelerated rearward. Since the occupant's velocity and energy is a squared function of acceleration time, the individual may be thrown violently toward the seatback with an impact involving the head or upper torso that could cause serious injury. What this means is that the position of the driver's head is critical for the proper functioning of the airbag (i.e. there is a 'safe zone' for the head), a fact almost unknown to the general public and certainly not appropriately communicated to the automobile driver.

The injuries may be gross, such as spinal cord paralysis, brain tissue laceration and necrosis, and can have consequences that may be immediately fatal, or else produce an abbreviated lifespan. However, the trauma may be not always be immediately recognizable in terms of overt neurological signs and symptoms and may not even be identified by image enhancers, psychometric instruments, or biochemical tests. They may be revealed by functional (not structural) imaging, sensitive neuropsychological tests, and sometimes by EEG pattern depression. This is because there seems to be another form of brain injury that is more subtle, not immediately clinically apparent or objectively measurable by routine testing. Time alone may reveal impaired cognitive functioning which leads to psychosocial difficulties in employment, family relationships, and in feelings of self-worth. There are individual differences in structural, functional, and biochemical response characteristics to be considered. In terms of the mechanisms of injury, in addition to the inflammatory processes, there may be a loss or alteration of synaptic gap or juncture connections (number, strength, remodeling, plasticity, recruitment, and function of neurotransmitter molecular proteins) (Peters, 1998a). The injuries may be diffuse from acceleration or focal from skull protrusions.

In some vehicles, the belted driver, in the absence of airbags, will have facial contact with the steering wheel at about 14 mph (22.4 km/h) sudden deceleration. Thus, the airbag 'must fire' deployment should be at or above 14 mph (22.4 km/h), since that may be the highest 'no injury' level. If there is a lower 'could (may) fire' or 'no fire' speed (such as 8 or 10 mph, 13 or 16 km/h), unnecessary injuries could occur from the airbag deployment. For example, with belted drivers, airbag deployment brain injuries have occurred at 12–16 mph (19.2–25.6 km/h) (delta V), spinal cord injuries at 15–18 mph (24–28.8 km/h), and quadraplegia at 10–12 mph (16–19.2 km/h) (NHTSA cases, 1999). Thus, contrary to general opinion, the injury problem is not strictly the result of high-speed collisions. It can occur from 'aggressive' (high-powered) airbag deployment at low collision speeds.

Thus, from a design standpoint, a less aggressive (depowered) airbag would be more appropriate for minor crash scenarios (low energy impacts). Design work is currently being accomplished on dual stage and variable-powered airbags, with sensors (e.g. infrared, ultrasonic, capacitance, or weight-sensitive) for better control of the airbag so it will function differently according to the actual risks involved. These, perhaps, would have a 'no-fire' in about half the current 'must fire' scenarios, the situations where harm could occur (children, small-stature adults, and occupants in the danger zone or out-of-position occupants). An integrated airbag system would also include belt pretensioners (for pulling the driver into a safe zone prior to airbag deployment), belt use (whether or not wearing belt restraints), occupant seat position, occupant weight and height, and crash severity sensors. Adjustable brake pedal and accelerator pedals should also be included as a 'universal design requirement', as well as a 'quality of life' objective (*Airbag Technology*, 1999; Peters 1997, 1998a; Peters and Peters 1984–1993; Yoganandan *et al.*, 1998).

Some early airbags had excessive permissible variation in bag pressure at critical moments (0 to 30 milliseconds) in the inflation time sequence, but such quality and product assurance problems have generally been resolved or improved as the technology has become better understood and refined in practice.

While awaiting design and quality improvements, there are health promotion communications that can reduce injuries in current-generation automobiles. The general public could be informed of the 'safe zone' requirements (namely that the driver's head and upper torso should be erect, the head at least 10 to 12 inches (25.4 to 30.5 cm) from the airbag, the head about 1 inch (2.54 cm) from the head restraint, the airbag aimed at the chin, recliner seats in the upright position, and the driver should be laterally centered on the seat and with regard to the head restraint). Since some seats have adjustment memory buttons for the convenience and comfort of the driver, one button position could be adjusted by the dealer to move the vehicle owner's head into the safe zone. This would not interfere with any adjustments needed for other drivers of the vehicle and would be an instructional lesson for the new purchaser as to the need for the head to be in a safe zone.

Side airbags and airbags for rollovers, knee bolsters, and head restraint enhancement are new safety features which may or may not need some health promotion; i.e. children should not position themselves under a rear side curtain airbag. Visual and audio warnings for out-of-position occupants are, basically, a design and marketing problem. The limits on many seat adjustments (forward and aft, up or down, tilt or non-tilt, pelvic support, etc.) is basically a design problem, assuming the vehicle purchaser is sufficiently informed about the selection of a vehicle that 'fits' in terms of safety as well as comfort. Health education could stress the importance of adjustable accelerator controls and brake pedals for those whose height may place them too close or too far from the airbags.

(d) Problems less amenable

At a 45 mph (72 km/h) rear impact, there should be no injury if the upper torso and head are properly located against the seatback and head restraint. However, many seatbacks (backrests) are weak and start to yield at about an 18 mph (29 km/h) rear impact, so the occupant may lift up or ramp up on the tilted seatback and suffer severe hyperextension or ejection injuries. This crashworthiness of a vehicle directly affects seat performance and injury severity (an illustrative test acceptance criterion is an acceleration pulse to the upper torso of less than 60–80 *g* for anything more than 3 to 5 milliseconds) (see Figure 10.2, head impact tolerance). There is little that can be communicated to the driver for injury reduction, but stiffer, stronger, or more rigid seatbacks have been designed into many newer vehicles. Some have a more horizontal displacement of the seat for better energy management. Some have low 'butt pockets' (lumbar deformable areas) for resisting lift-up or ramping, and pelvic restraints in the (bottom) seat cushion to resist 'submarining' forward under the instrument panel. Some seats are deeply contoured to resist lateral body displacement. Many of the newer designs are very impressive in their occupant protection effectiveness and even more substantial design effort is being expended to attain a yet higher level of safety in seat design. Photographs illustrating some of the issues discussed in this chapter are shown in Figures 8.4–8.17.

Figure 8.4 An adjustable head restraint is often left in the 'down' position, creating an unnecessary hazard or unsafe condition.

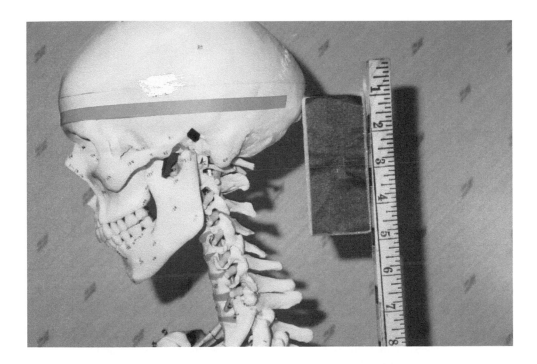

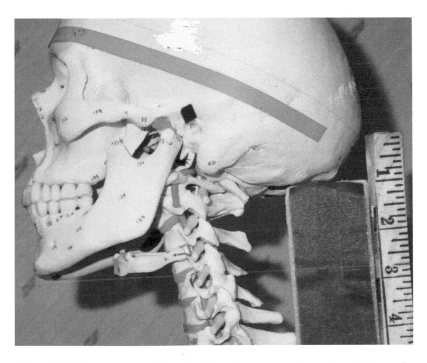

Figure 8.5 The curvature of the skull is an important determinant in hyperflexion, since an inch or two in the height of the head restraint may be critical if 'lift-up' occurs.

Figure 8.6 The driver may look to the rear before a lane-change maneuver in an emergency, crash-avoidance situation. As the driver tightens the arm muscles and leans forward, the chest is too close to the airbag. The use of belt pretensioners corrects this situation by pulling the upper torso back into an erect position before the airbag fires.

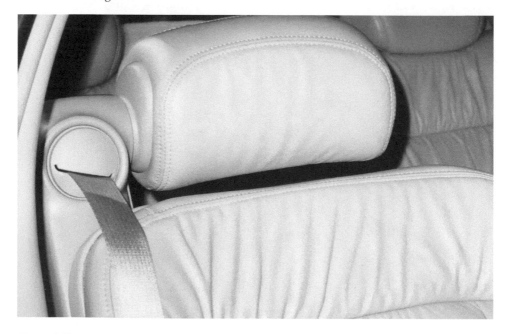

Figure 8.7 Integrated seatbelts (belts in seat) that go over the shoulder can reduce ramping (ejection to the rear compartment from reclined seats or high-yielding seats during a rear collision).

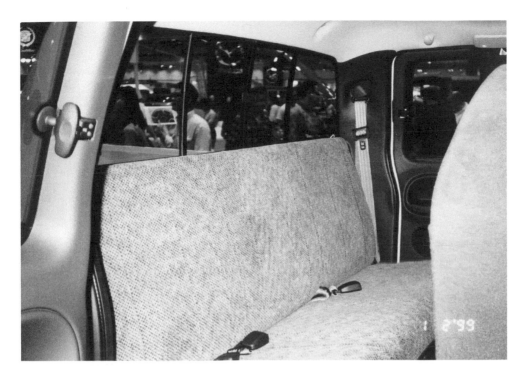

Figure 8.8 Some vehicles are sold with shoulder-height seatbacks and no head restraints.

Figure 8.9 Offset crashes require seat design considerations.

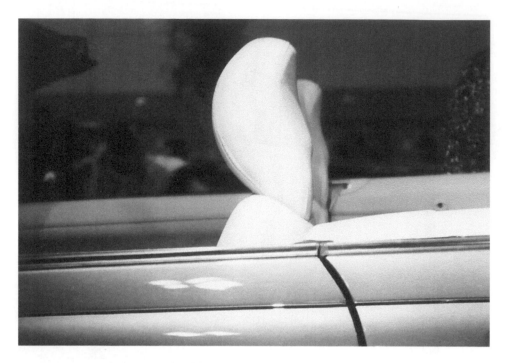

Figure 8.10 Head restraints should be about an inch (2.54 cm) from the occupant's head and reach to the top of the head.

Figure 8.11 What are the biokinetics when a driver leans over, looks at a display, and manipulates the controls? This is another form of out-of-position occupant protection problem.

Figure 8.12 Flip-down video monitors. Some vehicles have up to seven video displays, both flip-down and inserted in the rear of the seatbacks. In terms of a collision, what does this do to crashworthiness and occupant dynamics?

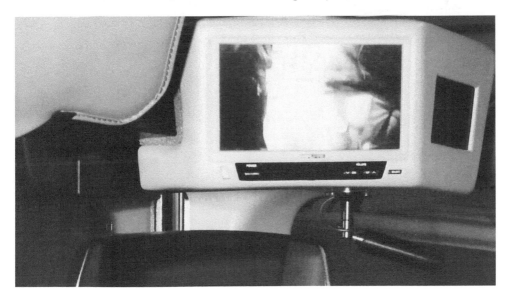

Figure 8.13 A three-way video display: one for the driver, one for the front seat occupant, and one for the rear seat passengers. It is located above the rear-view mirror. What are the occupant biokinetics with the head rotated to the right or left?

Figure 8.14 Body lifted to show the proximity of rear passenger heads to the rear window. Note the depth of the rear crush zone. With a mismatch-and-override rear collision, how does this affect occupant protection?

Figure 8.15 Four-point occupant restraints (two shoulder belts) have been used in specialty vehicles. Some automobiles are now scheduled for this major improvement in vehicle safety.

Figure 8.16 There is a new class of very small vehicles, used for both passenger and work functions. Some may be found on public roadways, used for parking meter (toll) coin collections. The occupant protection varies with brand and model, but is rather evident.

(e) Conclusions

In essence, motor vehicles are major contributors to unnecessary and unacceptable injury rates, but the automobile vehicle industry has a strong focus on design improvements that should be of great future benefit. However, there is a current need for greatly increased health protection efforts in order to communicate knowledge of existing hazards to the general public, to foster a better understanding of personal risk avoidance behavior, and to encourage safeguards that can reduce preventable harm. The injury costs are so high that there should be a cooperative and complementary effort by all those affected, including vehicle drivers and occupants, emergency and critical care specialists, lawyers and insurers, relevant government and crash test entities, highway and traffic control engineers, vehicle dealers and manufacturers, and others interested in preventive medicine, injury reduction and improved quality of life.

9 Human simulation applications

(a) Introduction

Simulation may not be new, but how the game is played has undergone substantial change. It is still rapidly evolving and has revolutionary promise in terms of fidelity, sensitivity, and in the wide scope of practical applications. Modern simulation has been made possible by dramatic increases in computer power and electronic systems capability, with concurrent reductions in time, cost, size, and complexity. Not to be overlooked are newly emergent needs, the forcing function of mandated compliance with specialized requirements and rather detailed specifications, and the availability of the talents of many highly skilled professionals. The promise is indeed great and the benefits are many, but there problems and pitfalls that should be recognized and overcome.

This chapter explores some of the consequences in the use of models that have not been validated, inadequately validated, have been underdeveloped, abused, or omitted. The focus is on digital human modeling as an exemplar within a systems context. The perspective is interdisciplinary, international, risk-based, and founded on the authors' personal experiences. Since terminology now varies, some definitions may be given to elucidate the actual concepts being presented. The primary dictionary definition of *to simulate* is 'to feign' or to make-believe or pretend. *Simulation* is defined as 'to assume the appearance without the reality' (*Websters New Collegiate Dictionary*, 1956). Thus, we often hear the admonition to remember that virtual reality is not reality, but it is just a first approximation that can save time, cost, market share, tactical battles, as well as human life.

(b) Early simulation

Years ago, a vehicle driving task was simulated by having a test subject sit in a portion of an automobile while a moving highway was projected on a screen. Thus, data was obtained on *human reactions*. We used a fixed cockpit simulator for research on aircraft instruments. Thus, we obtained data useful for the *design* of displays and controls. In that analog age, this was a cumbersome, costly, unreliable, and relatively inflexible approach.

A Link trainer, in the corner of a laboratory of an aircraft manufacturer, was composed of an aircraft-type cockpit mounted on articulating struts for limited physical movements in response to manual flight control inputs. The movements gave a little of the feel of actual flying, and were obviously only a gross simulation of the real thing. It was essentially a *training device* that gave the appearance without the safety problems of those who might not have adequate proficiency. It served to save time, cost, and lives. It attracted attention, emphasized the role of human factors in relation to the flying machine, and gave some limited research and development benefits.

For military weapons, simulation in real-life settings sometimes could be less obviously a pretense and could be highly instructive for design and development. But, overall, such research activities with humans have been severely restricted in terms of controlled simulation.

Today, computers are everywhere, processing enormous amounts of information, and are low-cost, small, and much more user-friendly. The reliability and usefulness of almost all instrumentation is remarkable. Simulation has come of age and more complex human research can be conducted.

The early paper-and-pencil human simulation has taught us much. We conducted reliability engineering at the component level, but in terms of human factors the foundational data were often lacking or just not relevant. To achieve system engineering objectives and product development goals, the human factors' estimates were sometimes quantitative, sometimes qualitative, sometimes gross, and sometimes speculative assumptions pending further research or data acquisition. Fortunately, these predictive and analytic simulation processes could be subsequently checked or verified in the design process by simulating actual operation with various assemblies, equipment, and systems. This laboratory demonstration, or on-site testing simulated reality before the ultimate test: the actual functioning of the equipment in the use-environment in such modalities as installation, operation, servicing, maintenance, repair, recall, and recycling.

All of this should indicate the breadth of scope of possible applications of human figure modeling. Of course, with the adoption of universal design, there are many new human attributes and variables deserving research attention (Peters and Peters, 2000a). As we focus on human figure modeling and relevant verification and validation issues, it is the actual usefulness and consequences of simulation that deserve special attention.

(c) Benefits and limitations

Simulation has been encouraged as a means of saving money and shortening design time, discovering and resolving problems before marketing goods and services, and preventing harm to test subjects. But there is much more to it than that.

(1) Inappropriate validation

A major automobile assembler had accumulated extensive crash test data from low-speed front barrier tests on earlier preproduction models. To save the costs of crashing new model vehicles, it developed a computer simulator crash model to attain the desired data from an equivalent to actual testing.

The simulation model predicted the biokinetics of a 50 percentile male, the head impact force vector, and it used a pass–fail criterion of g-forces where a 20% level of serious injury or death was acceptable. Unfortunately, in a real-life crash, a 96 percentile male's head did move forward but then rotated downward, so the impact on the steering wheel was at a point of its greatest hoop rigidity and its pathway was through the driver's mouth up to the floor of the brain. Thus, the injury prediction did not include a full range of drivers in terms of anthropometry, had a tolerable level of injury that was questionable, and failed to predict appropriately the biomechanics of injury. The critical crash pulse differed from the model because of front-end structural crush zone or progressive yield differences. Movements of the steering wheel during the time of head impact were not considered. In essence, there were too many unsophisticated assumptions and inappropriate validation criteria. Thus, real-life crashes that should not have caused injury resulted in serious permanent injury. Therefore, merely collecting accurate data is not enough, since simulation should adequately predict the ultimate real-life criteria, entail relevant correct assumptions, and not tolerate risks that are legally, morally, and ethically unacceptable.

(2) *The no-test approach*

Where there is perceived low risk, a common philosophy in *design* is a cut-and-patch development and in *manufacturing* it is 'ship it now and fix it later'. The absence of testing is often accompanied by some very simplistic assumptions about human performance. For example, a power tool had many internal part failures during customer use that could have been discovered by factory testing. The marketplace was tolerant to a credit and repair effort, so there was no recall program. There was also no human use testing, only assumptions about normal use, so there was some surprise when serious injuries occurred during use. In essence, the product specifications did not include appropriate material testing nor was there any simulation of customer use during normal operation or during parts failure. The lack of preproduction simulation was costly in terms of money and market share.

In another example of a failure to test appropriately or conduct a simulation study, there was a series of fires in mobile homes or manufactured housing in America that killed many children. We recall one such incident in which a mother was standing outside her mobile home, about 100 feet away, when she saw a fire erupt in the middle of her home. She quickly attempted a rescue of her children, but was driven back by the toxic smoke and hot fire that spread explosively fast within the home. Rescuers were unable to extricate the children from the rear of the home because of a mismatch between the size of the window openings and the rescuers' human dimensions. The children's bodies were found later in the usual location, at the rear under a bed. This example illustrates the human 'dimensions-plus' problem: it is important to study the human *reactions* and response *processes* to predictable dangers, emergencies, risk mitigation, escape, and rescue scenarios. That is, human dimensional studies should be supplemented with human performance studies, in a process and system context, with time, movement, warnings, and other factors included.

It is of interest, in this example, that the mobile home manufacturer sold a fire-resistant model, with a similar appearance, at the same time that the accident model was sold. But the family were unaware of this fact when they purchased their mobile home and moved it into an attractive residential park. The availability of a fire-resistant model suggests some constructive or actual notice of the life safety problem by the manufacturer and a partial design response to it. However, fireworthiness is good, but survivability is better. Appropriate simulation could have resolved the entire problem, but how should this be accomplished? It would be beneficial to formulate a voluntary standard or *recommended practice guide on human simulation* that could be used by large or small companies in many industries.

Incidentally, the mobile home fire danger problem illustrates why we need data on children and the elderly in human figure modeling, human responses to danger, and the biomechanics of escape from danger. Thus, there is a real need for extensive research data to support future simulation models that could significantly reduce risk and liability.

(3) Late testing

Some years ago, an automotive vehicle failed certain road crash simulation tests, but the testing occurred too late in the production cycle for any changes to be made before marketing. One major injury-producing problem was corrected, but not until two years later. The difficulties could have been avoided by early computer-aided design of the vehicle body and platform, plus computer-generated graphic human digital modeling. Today, simulation can and should occur early enough to correct major mistakes, to enable time compression of the design process, to facilitate better or more creative concept design, and to minimize detail development mistakes.

(4) Human predictability

In simulation, there are generally three basic theories on human performance as part of a system, product, or service. First, the so-called uncertainty and *unpredictability* of human response, a belief that is legend, and which may be true if based on uninformed or speculative opinion. This theoretical approach may help rationalize poor design decisions, but is a rather hopeless approach in terms of achieving good design or avoiding legal fault. Second, the so-called logical predictive pathway, based on the assumption that people *behave rationally* and are primarily influenced by instruction, habit, and self-interest. Of course, people do not always act rationally or as might be expected logically. This theoretical approach offers only marginal utility because it is so subjective. Third, the so-called *test-and-find-out theory*, using representative populations on representative equipment. The issues to be tested may evolve from analytic techniques and be available from empirical human error data. It requires realistic simulation exercises and confirmation by eventual field data. The approach is to predict, test, and find out what really happens in a given situation. In essence, we need 'real data by real testing' for determining foreseeability and the timely correction of induced human error and undesired human performance.

(5) Sign-off assurance

Those who participate in formal design reviews may receive a bundle of engineering drawings shortly before the review. Time is limited and there is a need to visualize in three dimensions from two-dimensional drawings, attempt to integrate disparate drawings, and detect errors and omissions. Anything that would help the visualization process would be welcome. Simulation graphics could help if timely available. Similarly, those who sign-off on a drawing would have more personal assurance if there were simulation graphics. The problem of visualization has led to the conventional mockups and physical prototypes which often reveal that there is not enough room for feet or shoulders, an excessive reach distance to manipulate controls, knee contact with hard objects, or line-of-sight problems. We are familiar with *new* products that still have such adverse human interactions and dimensional problems. It is suggested that, in the design cycle, simulation be applied early, mockup later, and prototypes last in order to save time and cost.

(6) Controlled incremental simulator research

There are some forms of controlled experimental research best accomplished on a realistic simulator. For example, there is a driver distraction problem with vehicle telematics, which involves increasing the load on the prefrontal working memory (that part of the brain that deals with reasoning and choice behavior). What is the effect on human visual perception or selective attention as the cognitive load increases? What is the capture efficiency of roadway hazards under vigilance, tracking, and crash avoidance situations? We understand that there may be three stages as the load increases: first, a gradual cognitive tunneling (narrowing of the focus by suppressing peripheral information); second, a more generalized (peripheral and central) perceptual degradation as inadequate brain filtering of the visual field permits non-relevant information to interfere with recognition and response time; and third, an overload of the working memory with brain circuit shutdowns that cause confusion, error, and disruption of relevant cognitive processes. Thus, simulators could gradually increase the mental load and determine both linear and non-linear effects. Is it true that peripheral vision is lost first, then a general degradation occurs, and finally a drastic loss of overall visual recognition? How much of a load is reasonably permissible from a distractor acting singly or independently, from distractors acting in a cumulative additive fashion, and distractors affecting the driver in a multiplicative manner? Such questions seem best answered by simulator research.

(d) Consequences

(1) Inadequate validation

The use of models that have not been adequately validated may have unexpected consequences. Failure to appropriately verify and validate simulation processes and data may result in user discomfort, violation of consumer expectations, personal injury, property damage, impaired system effectiveness, or loss of mission objectives.

A few simple field examples may illustrate the point that serious consequences could result from assumptions regarding human figure models. Most people are now aware that short elderly females should not sit too close to the steering wheel of automobiles that contain airbags. We really learned this lesson when we reviewed a minor automobile collision in which the driver suffered a burst aorta. We checked a book we edited in 1991 and were reminded that the first-generation airbags were designed to protect the average male occupant in a 30 mph front barrier accident. The design approach (Breed and Castelli, 1991) was to optimize the system for a 50-percentile male and, hopefully, extrapolate for the 95-percentile male to the 5-percentile female. Design engineers were warned about small females, out-of-position drivers, and the need to consider the spectrum of the real world. But, subsequently, there were serious injuries and deaths. Then there were more explicit warnings to drivers and development of adjustable brake and accelerator pedals (for short legs, to help achieve proper eye level, and to gain distance from the steering wheel). Then there were developments in seat sensors (for occupant size and location) and variable powered airbags (also factoring in collision forces). That is, there has been a known human figure problem from the start, but adverse consequences over a 10-year period of time, and finally some good concept design, much of which is now in the testing stages.

(2) How much validation?

Of course, we now know that children should be in the rear seat of automobiles because of design problems. That is, there was a failure to include children properly in the design and simulation process. The simulation process should include 'what if' tools and fault coverage to enhance the likelihood of field success. How much incremental verification and ultimate validation is needed? How much is enough for quality and product assurance? We will discuss such problems as we proceed.

(3) Human size problems

A friend, an average height male, has a heavy upper torso and selects luxury vehicles and SUVs for personal comfort. He tilts the reclining seat backwards to gain room and comfort. He was reminded that the owner's manual stated that the seatback should be kept upright when traveling and was told that weak tilted seatbacks permitted ramping upward and backward in rear-end collisions. Although made aware of the risk of injury, he still preferred the benefits of comfort not otherwise achievable for someone with his body dimensions.

We asked a machine operator to re-enact how he was seriously injured on a large plastic injection molding machine. There was a barrier guard or large gate covering the danger area where two parts of the mold came together. It seemed appropriate for an average-sized person who needed to reach under the gate to remove formed pieces. However, the small statured machine operator could actually sit on a ledge, under the gate, and reach beneath the danger area believing it was safe. An articulating arm of the mold caught and trapped his chest, there were head and brain injuries, and he was pinned by the machine until the molds could be manually moved apart. The barrier guard may have been adequate for a 50-percentile person, but that area of the machine was a trap for a short person. We have seen plastic blow molding machines where tall and long-armed persons could reach over the guard to wipe away excess plastic and not stop the production process to clean-off the molds.

There are swimming pools with bottom depth dimensions adequate for plump swimmers, but dangerous for lean swimmers because of the trajectory of divers in the water.

There have been medical device and implant problems because of human sizing. There have been human-presented area or projected human surface area problems for police and military weapons.

In essence, there have been an endless list of adverse consequences from inadequate validation of various designs and simulations involving human body dimensions and kinematics.

(4) Billion dollar incentives

According to a 2001 Securities & Exchange Commission filing, one automobile vehicle assembler had damage claims for just one tire safety problem of $590 million, plus $603 million for an occupant restraint system problem, $1.7 billion for asbestos claims, and $2.4 billion for another vehicle rollover defect. Overall, the contingent liability was $4.7 billion (US). Obviously, this legal liability seems excessive and appropriate proactive simulation seems to have great potential in reducing the magnitude of such claims.

(5) Simulation to shorten design time

Computer-driven animated anthropometric models have been used to virtually explore and define human interaction in the use, maintenance, and repair of products or systems. This may seem to be like a video game or an entertainment virtual reality. The game of pretend has become more and more real, with refined immersive content such as higher resolution imaging, minimum reaction latency in interactions with objects, audio or sound effects, and even some form of physical motion sensation or experience.

When the simulation feels right, it just seems more credible, accurate, and complete. Thus, there is a great temptation to go from a complete model or virtual reality directly into final design and even directly into manufacture. This skips over much of detail design, the clay models and prototypes, and many tests. Obviously, it increases the speed from concept design to production, but does not allow room for errors, defects, and bugs that normally appear and are resolved in the product development cycle. This may not be a problem where there is great flexibility for in-field or on-site design improvement. But this is not a perfect world and simulation should not be extended beyond its inherent capability. We have seen simulation applications that create dormant problems that become manifest some years after the production run.

Simulation can shorten design time, but some form of normal product or system assurance is usually needed to find and correct the errors that historically appear in a non-perfect world.

We suggest caution in the generalization of simulator-derived information, that causation be scrutinized, and conclusions be relevant and to the point. System implications should not rest entirely on the seeming authority of a simulator. Critical thinking is still very important.

(e) The only reasonable test

(1) EMC and health effects

In the design of automotive vehicles, we are increasingly concerned about electromagnetic compatibility (EMC), radiation susceptibility, and electrical field effects. The interferences may be deep within the electrical devices and computer microchips, in the signal pathways and wiring harnesses, in the reception and transmission of cell phones, or in the performance of navigators. We understand that interior vehicle body resonance may amplify the adverse effects, perhaps tripling the electromagnetic field (De Leo *et al.*, 2000). In addition to the external fields and the resonance effects, the electronic devices themselves, within the interior of the vehicle, can also enhance the electromagnetic fields.

While we may be concerned with the performance of an in-vehicle cell phone, we also should be concerned with human EMC. Are there long-term health effects? If so, where should the electronic devices be located or shielded relative to the human occupants? What are the effects on persons of varying dimensions or body shapes?

The issue is the use of simulation, rather than human subjects, in identifying the strength of electromagnetic fields and their possible effects on human cells, tissues, and organs. Simulation seems to be the only reasonable approach.

This may also be similar to human performance under the effect of some combination of multiple distractors. Are the cumulative effects additive or multiplicative? Unless we test we do not know.

The build-to-order concept and that of customizing vehicles suggests that a case-by-case approach for the electronically loaded vehicle is far too expensive. Predictive simulation models, for that situation, may need to be developed to assure good equipment and safe human performance.

What may be the fault consequences of the use of only simulated electromagnetic fields on the performance of physical equipment, without using digital human models to determine possible health effects and alternative designs?

(2) *The debugging process*

We have entered an age of computer-aided design, manufacture, and product control. Thus, computer-generated human digital modeling is or should be compatible with these new processes. But the design process often involves many changes to cure unexpected interaction effects, added features, improvements to attain desired performance, and constant debugging. Design reviews produce changes. Changes to improve manufacturability also result in design changes. The problem is that each change may affect human interaction. Simulation models provide quick assurance that human performance has not been compromised during debugging and change. We recall one aircraft ejection seat that jammed halfway out of the cockpit because of a manufacturing change that was not communicated to the design engineering group. All substantive changes need checking by simulation even after release to manufacturing.

(3) *Tuning for the system*

With electronic devices, there may be an instruction-set simulator that permits tuning a device to meet system requirements, to identify worst-case situations, and to determine best utilization under the circumstances (Ohr, 2001). Similarly, human task optimization could be achieved by small-scale simulation of operation, servicing, maintenance, or repair tasks. In terms of business development plans and military war gaming, appropriate simulated task performance on a large scale may reveal real-time responses to unexpected situations and preemptive interruptions. Small-scale simulation to fine-tune a set of tasks on a prioritized list may be the only reasonable test available. Large-scale simulation for financial, technical, political, and military processes may also be the only reasonable testing available. There are many historical examples, industrial and military, of the tragic results of poor decision-making based on subjective opinions and lack of appropriate data. That is, industrial and military models that were not validated brought forth disastrous consequences. Simulation permits informed decisions based on objective data, the testing of alternative hypotheses, and the use of risk-tested scenarios. In essence, simulation can be used to fine-tune a very small or a very large system to gain the advantage necessary for success and it may be the only reasonable test.

(4) *Informed consent*

Human simulation could help to avoid serious ethical problems where humans are now used for research purposes. As has been well publicized, some major American universities have had their research suspended for ethical violations dealing with the protection of research subjects. One problem is appropriate informed consent and another is the actual functioning or effect of the institutional review boards or their equivalent oversight entities. Since it is a growing problem, in both medical research and engineering development, future compliance may require creative and alternative research procedures of which human simulation is one possible choice. Simulation could help to eliminate some of the hazard exposure, subject selection, and ethical monitoring problems particularly in early experimentation and in engineering concept development activities.

In terms of identifying critical human behavior modes, for use in the development of human simulation models, techniques such as focus groups, interviews, community or on-site reviews, and factory demonstration or useability studies may not be sufficient. The use of unintrusive monitoring by camera or observer does present some ethical problems. Such data-gathering, in the support of model development, will probably become more intrusive and detailed as graphic images become more realistic. We can expect both the perceived realism and the breath of applications to escalate quickly to high levels because of the use of the free Linux operating system and its open-source code improvements, low-cost high-performance hardware and work stations, the pressure of digital-content demands, and the growing sophistication of experienced animators. Life-like animation certainly will be more believable, credible, and seemingly true at a time when it may be very speculative and even false as to human capability and foreseeable human reactions.

(5) Military simulation

We think of battlefield simulation in terms of feigned combat maneuvers. It acts as a troop training exercise, an inter-unit and command hierarchy communications test, an evaluation of command and control theory or doctrine, systems development, and the utilization of emerging technology. Certainly, this is preferable to small or large wars, with cost and casualty consequences.

It is often large-scale simulation with many objectives and benefits. It may be quite realistic to those involved or not realistic at all.

In terms of information simulation, the criteria might be availability when and where needed, its authenticity and integrity, and its ease of use. The ultimate criterion is the effectiveness of the resultant human decision-making at all levels of the military hierarchy.

Simulation is also used to evaluate coalition or multinational operations, crisis response, and political–military interactions. There are large-scale simulations of considerable importance, but so are some private business risk analyses (Peters, 1997) that involve simulation. Modern business is fundamentally competitive or warlike, dealing with product or economic threats, the risks of marketplace attacks, and the acquisition and manipulation of information in intellectual warfare. So large-scale simulation in business activities is to be expected with very large financial incentives for those who win their business wars.

We sometimes forget that the military is an aggregation of interacting humans, in novel settings with human factors problems everywhere. In the 1950s, there were many human factors research projects that provided everyone with basic guidance documents that are still in use. The early emphasis was on human dimensional data and human figure modeling to provide proper fit, access, and biodynamics. This was for design improvements in clothing, aircraft, ships, trucks, tanks, personnel carriers, facilities, etc. Of course, the basic populations have undergone dimensional changes over time. The derivative human models have significantly improved with the advent of solid-state electronics, computers with high capability, versatility, low cost, and availability. We find human figure modeling and biomechanics graphics in the forensic area, models with a type of virtual reality applied to early engineering design and manufacturing, and dimensional models being used for a variety of purposes. It is surprising to see the wide scope of applications and the extension to other human factors variables and attributes.

Simulation is now essential if *universal design* objectives are to be meaningfully accomplished (Peters and Peters, 2000a). In terms of human dimensions, the automobile driver now has adjustable seats, adjustable head restraints, adjustable occupant restraints, brake and acceleration pedal extensions, and other *compensatory features*. There are designs where there are automatic adjustments, including even of the instrument panel and steering wheel, to achieve proper eye position. Some cultures may include special provisions for hats and headdress. This illustrates the human dimensional needs and applications. Consider the fact that in some cultures, clothing such as men's shirts is sold in closed containers without the assistance of trained salespersons, with the purchaser's actual fit unknown. To what degree is carpal tunnel syndrome related to lack of application of universal design principles and human dimensional limitations? We see industrial machinery where machine function alone almost completely dictates operation and control location, the human dimensional requirements, and the inherent accident risk.

(6) The many forms of simulation

The use of mathematical models to approximate a real-world situation involves the use of abstract symbols in a fairly remote context. It may be a very useful first approximation of reality or helpful in preparing for more realistic simulation. In terms of reliability engineering or system safety engineering, these computer-aided evaluations have been helpful in design engineering to the degree that real-life qualitative and quantitative data has been available. There are many system effectiveness and risk assessment models that approximate or simulate real-world problems.

(f) Intentional bias

(1) The gold-plated component

The 'gold-plated component' is a specially designed and fabricated component that can pass all tests and achieve customer satisfaction at an early stage in engineering development when a production version might fail. An illustration of a gold-plated human component is the experienced truck driver who can maintain control of a vehicle, after a front tire blowout, with little deviation from a straight line. A typical truck driver who does not expect a blowout probably would lose control. It is also akin to the automobile partial manikin sled test easily arranged to yield some desired, but misleading, result. It is also similar to a group of specially trained and experienced soldiers who can maintain order and direction in a mock exercise that would confuse most other soldiers. The gold-plated component usually defeats appropriate simulation verification and validation.

(2) Liability defense

Good graphic simulation models should be welcomed by defense counsel to help prove in court that all reasonable measures were undertaken to achieve a high level of safety. Design graphics are better than words and such demonstrative evidence generally has good jury appeal. Another form of simulation is the legal *mock* trial and much of its value is in the identification of representative jurors, their characteristics and biases, and their responses to alternative forms of communication. It helps to weed-out the 'bad user' and the false assumptions of the attorney–experimenter.

(g) Six caveats

1. We have to move vigorously and expeditiously from the *art* of animation to the *science* of human modeling.
2. We can expect considerable *progress* in graphics, just as radiographic imaging has been quickly supplemented by fMRI and many other imaging techniques that even permit imaging of gene protein expression. Similarly, human simulation is also a technique capable of rapid development. In addition, we understand that the simulation of molecular structure is now an *effective tool* used in pharmaceutical research. Quite simply, human simulation is a similar tool or technique that could present unusual opportunities for creative professional application.
3. Those of us who have reviewed hundreds of crash test videos to determine human biodynamic or biokinetic responses realize the importance of accurately converting to simulation models that can be better visualized. If one objective is to permit *better visualization* of a condition or process, it suggests the importance of validation to ensure that the correct target is kept in sight during the development of the model. Of course, the ultimate objective is accuracy of the technique, *appropriate perceptual and conceptual interpretations or visualization*, and productive applications based on needs-based research. For example, we recently reviewed some medical research where the graphic simulation had been simplified, but the omission of details may have seriously misled the research investigators. In essence, what is irrelevant or clarifying to a model builder may create unnecessary problems for the user. Thus, research is needed to assure the accuracy of the technique, its proper visualization, and to enhance potential applications.
4. Human dimensional models could include both the exterior human surfaces and the *interior surfaces*, such as sizes of airways and other passageways, locations of force generation and effects, and spaces available for biocompatable implants. There are many potential applications of human simulation, such as a teaching tool for complex subjects and the use of bioimagery to identify decision choices and research hypotheses.
5. It should be understood that there can be *legal fault* for defective models. This includes liability founded on legal theories of strict product liability, the negligent failure to exercise reasonable care, and for improper claims and representations resting on theories of warranty and misrepresentation. Contingent liability may be difficult to assess because legal fault may vary as to locality, time, court procedures, judges, the political climate, and inter-company relationships. However, it should be clearly recognized that the magnitude of damages may be high because complete product lines could suffer from a failure to validate the model properly.

6. The potential benefits of human simulation may be realized by any country, company, or endeavor provided that they properly plan for the needed research. In a *competitive world*, those who proactively conduct the necessary research will have the advantage in scientific momentum, patents and intellectual property rights, product innovations, process derivations, and trade exploitation.

It is now possible that a person possessing an excellent working knowledge of human factors and ergonomics can position and move computer generated hominoids to predict select human performance capabilities of groups of people within a computer rendered environment.

(Don B. Chaffin, University of Michigan, personal communication, 2001)

The single most serious weakness in the human performance modeling field is the lack of robust data bases built on observations of human performance.

(Name withheld by request, personal communication, 2001)

(h) Definitions

(1) Verification

Verification (accuracy) formally authenticates or confirms that the operation or process does, indeed, accurately and reliably measure that which it purports to directly measure.

(2) Validation

Validation (relevancy) assesses or confirms the quality or effectiveness of the model in terms of its usefulness or ability to predict behavior or events in terms of an ultimate criterion, practicality, or reality.

(3) Sensitivity

Sensitivity (measurability) may be considered as simply the *power* of a measurement system. We define sensitivity as the ability to identify, measure, and discriminate a particular desired variable, with a sufficient range of measurements that are reasonably precise and free from extraneous confounding, and with the capability of interpretations that have a useful or meaningful result.

(4) Relevancy

Relevancy (usefulness) is a measured and interpreted value that may be evaluated or utilized by comparing it to an arbitrary or reference criterion. For example, g-forces and their time duration from simulated persons (that is, manikins or dummies used in crash testing) may be compared with the head impact criterion (HIC), other acceptability limits, or a scale of estimated injury. In other words, what is measured should be related or relevant in a practical, useful, or problem-solving sense.

(5) Simulation – legal

While the word 'simulation' may have some favorable connotations for those of us who have utilized the technique or methodology, we should be aware of other less favorable definitions of the term. In French law, it may mean collusion to present a fraudulent deceptive appearance. In the United States, the law may define it as a pretense, concealment, or counterfeit activity that may be attacked or disregarded (Black, 1968).

(i) Workplace applications

In terms of occupational injury prevention, the human range of motion data should include something more than human capability. The limits should include reasonable injury extrapolations (static and dynamic), based on detailed human anatomy, tissue interactions, modern neuroscience, and, eventually, cellular considerations. This is necessary for adequate injury prevention, such as avoidance of torn ligaments and fenestrated tendons, fractures and displacements, neuritis and neuromas, and carpal tunnel compression neuropathies (including ulnar and radial, as well as the median nerve branch). This suggests that simulation as a tool or technique can have independent merit and that validation criteria can be critically important, sophisticated, and motivating. In fact, properly conceived validation criteria are the keys for such advanced concepts of human simulation. To make safety acceptable, it could be subsumed under human performance.

(j) Conclusions

There are some guidance principles that are worth remembering, such as the following.

1. Computer-generated digital animation is emerging from its infancy, whether animated actors or synthespians, virtual announcers or newscasters, or the computer imagery in forensic accident reconstruction and engineering design. As the technology improves, *the application will multiply*.
2. Computer characters, however lifelike in appearance and movement, need a script. The script requires human performance data and extrapolations that can only come from a serious research endeavor. The appropriate technology may become available, but *its effective use may be limited* by insufficient human-derived data proven valid for a particular application.
3. The computer model may be measurably accurate, realistic, and verifiable, but invalid or poor in predicting a desired objective. That is, a seemingly good test may not be relevant to a desired injury-prevention goal. *Validation is the key* to successful models.
4. Human figure dimensional and movement modeling is the *vital first step* that should lead to human performance, human reaction, and foreseeable behavior models and animation.
5. The virtual reality found in simulation is not reality. The approximation is as worthwhile as the *worth of the input data*, the methods utilized, and the validation process.
6. The cost and consequences of simulation mistakes and omissions can be very high. This is particularly true where there is unjustified over-reliance on this form of

testing or where there is legal fault for failure to take appropriate responsibility in terms of validation.

7. Graphic simulation permits rapid and creative prototyping which speeds up the design cycle, but this is only early concept exploration and error-reduction. *It is not a substitute* for subsequent testing before product release to the customer. Skipping the final testing is gambling in a situation where history generally indicates there are poor odds in culling out all possible design mistakes.

8. In human simulation, the emphasis should be on human performance in a *systems and process context*, not just in components and static structures. It should always be keyed or referenced to ultimate use, service, and disposal operations.

9. Graphic computer-generated versions of accident reconstructions can be illustrative of the degree to which good visual and auditory communications can have a *dramatic effect* on a targeted audience such as a jury, corporate management, or prospective customers. But those graphics may reflect to some degree a form of an innocent but overzealous, very enthusiastic, overly optimistic, and unrecognized overkill. Always look critically at what is being presented, its purpose, and the actual value of the results.

10. More serious is simulation conducted with premeditated *bias*, perhaps open advocacy, and by the intentional selection of test procedures executed to produce a desired result. Instilling bias may be an attempt to convince others that all design requirements have been attained or that some hypothesis is true. Thus, for any simulation there could be prejudicially biased results and caution should be exercised before detrimental reliance occurs. What seems convincing may not be true and confirmation is always helpful for informed decision-making.

In conclusion, simulation, as a technique, should have a powerful catalytic effect on human factors research, application, and practice. It could serve to promote coordinated, beneficial, interdisciplinary design efforts. More importantly, it can significantly improve design, lower costs, and save lives. It is a commitment to progress that deserves substantial intellectual effort and appropriate financial investment.

> Requirements have necessitated devising such varied approaches to human simulation as mathematical models, two and three dimensional models, partial simulations, animal subjects, and voluntary as well as unintentional human exposures.
>
> (Severy, 1968)

> Each technique of automobile collision simulation, whether physical or analytical, has its own limitations and difficulties.
>
> (McHenry and Nalb, 1966)

(k) Relevant terminology

(1) Snubber (component simulation)

A full test vehicle or partial vehicle body may be moved at a constant velocity, then quickly decelerated by cable, pneumatic, or hydraulic mechanisms that can provide predictable and repeatable decelerative forces on the object being tested in or on the vehicle. The test vehicle is not destroyed in the process, but the available travel (distance) is generally very short. The term 'snubber' applies to test devices that are *snubbed*, slowed, checked, or stopped to achieve a desired deceleration force on the test object.

For example, an upper torso manikin, with or without occupant restraints, may be decelerated at given deceleration pulses to determine differences in head motion and the location of possible impact. The components being tested simulate those that will be in actual use by the customer.

(2) Impact sled (component simulation)

The test vehicle or sled may be accelerated, to meet a time and acceleration curve (waveform or crash pulse) rather than be decelerated to meet the test requirements. The test vehicle may be larger and the distance traveled may be further than that of a snubber. A reversal of the vehicle can simulate rear collision forces and a 90-degree rotation could simulate side collisions.

(3) Static and dynamic tests (component simulation)

Bench-type static testing and dynamic (moving load) testing may be performed, as component testing, on seatbelts, door latches, and other items. The test results may then be verified by use of snubbers, impact sleds, and barrier testing usually in a system context or product application. It is less expensive to simulate performance, strength, durability, and weathering at the bench level to determine compliance or make improvements before going on to more expensive testing.

(4) Barrier tests (impact crash simulation)

Automotive vehicles may be impacted into barriers, of various types and manner of body engagements, to determine the actual effects of a collision. These are costly not only in terms of instrumentation and manpower, but also in terms of the loss of damaged vehicles. Other simulated testing would normally precede barrier testing. Barrier testing simulates real-world crash impacts under controlled conditions.

(5) Test track (vehicle system simulation)

Test track, test center, or proving ground testing can be in the form of accelerated stress testing to determine vehicle system integrity, durability, and the effects of customer usage profiles. For example, a 100,000 mile (160,900 km) criterion may be used to identify fatigue, failure, durability, and customer satisfaction, but the equivalent test track driving distance may be only 9300 miles (15,000 km). Such testing may be outsourced either for costs or for independent testing credibility. The test attempts to simulate actual customer use of the vehicle.

(6) Human digital modeling (human simulation)

The use of unembalmed human cadavers to represent live vehicle occupants during vehicle testing has been largely discontinued. The use of physical manikins is now common, since they can be well instrumented even if they are not true to what is desired in terms of biokinetics. To reduce costs and increase useful applications, attempts are still being made to combine test data for optional computer models that can be adjusted for new vehicle models and to outsource supplemental testing on unique design features.

10 Crash testing

(a) Human testing

(1) Introduction

One very important objective of crash testing is to determine whether an adequate level of occupant safety has been achieved. Obviously, live humans cannot be sacrificed in any kind of experimental (research) or development (engineering) testing if human subjects could be seriously injured or killed. In fact, as it now stands, the first actual live human testing or ultimate validation of the design takes place when a consumer or user experiences a real-life vehicle crash or collision. This fact emphasizes the importance of collecting precise, accurate, and relevant on-site accident reconstruction data, independent of any liability defense effort or government regulatory or involuntary recall concern. It also suggests the real importance of human simulation.

(2) Volunteer testing

Most of the early testing with humans occurred many years ago when engineers, inventors, stunt drivers, scientists, and automobile enthusiasts used themselves as human guinea pigs or volunteer test subjects. The personal injury risks might be relatively low for the low-speed horseless carriages of 100 years ago. As vehicles gradually evolved into high-speed sophisticated automobiles, the risks became too high for a voluntary informed consent. In Europe and elsewhere, race drivers took high risks to become famous. In the United States, carefully premeditated self-sacrifices reached the zenith with the decelerative sled tests of Colonel John B. Stapp, USAF, which produced much useful human tolerance data.

Generally, when human volunteers were used for the collection of impact tolerance data, it involved a gradual increase in the force until the subject had subjective discomfort (the 'ouch' or 'please stop' criterion) or when the research scientist believed that a significant risk of injury could occur to the subject. In one test, human volunteers were exposed to *increasing* levels of sled acceleration and velocity (Hendler *et al.*, 1974). In another test, the forces were *kept below* that which was believed to cause injury. The human volunteers (subjects) were exposed to low accelerations (6 g and 9 g) on sled tests (Ewing *et al.*, 1975).

The sometimes heroic and sometimes foolish attempts to contribute to automotive engineering could not continue as information accumulated on injury causation. It was not just the obvious life-threatening injuries that were of concern, but the previously unnoticed and somewhat unpredictable minor injuries that could result in permanent physical and mental impairment. There were unanticipated and unacceptable risks in conducting human tolerance research. There were too many delayed-reaction injuries. A new class of brain injuries emerged with the advent of more precise imaging systems, sophisticated neuropsychological testing, and modern neuroscience.

The question became how to obtain data applicable to human responses when live human subjects could not be tested in a dynamic vehicle situation. There were studies involving the use of animals, but they were too different in structure to provide the relatively precise and useful data needed.

> All proposals to conduct research involving human subjects must be submitted for review and approval to one or more *independent* ethical and scientific review committees. [Emphasis added.]
> (*International Guidelines for Ethical Review of Epidemiological Studies* (1991), Council for International Organizations of Medical Sciences, Geneva)

(3) Cadaver testing

There have been attempts to gather human data by using dead people (cadavers) in crash tests, by dropping them head first down elevator shafts, and for other biomechanical testing on body structures, organs, and tissues. Some good general data and basic concepts emerged, but even fresh unembalmed cadavers do not have the muscle reactions or other tissue responses of live humans. There were also informed consent, moral, and ethical problems in damaging or destroying the corpses of the unclaimed bodies of the adult homeless, unknown older people, or youthful orphans.

During the period in which cadaver testing became an alternative to human testing, there were many studies that might not be acceptable today. For example, blunt thoracic impact experiments were conducted with unembalmed cadavers (Nahum *et al.*, 1975). There were studies of the biodynamic responses of embalmed cadavers, with their abdominal organs eviscerated and strain gages applied to their vertebral bodies (Begeman and King, 1975). Impact tests were conducted on the knees of seated cadavers using an impact pendulum (Powell *et al.*, 1974). There were studies of the mechanical properties of bone under conditions of impact, bending, torsion, shearing, tension, compression, fatigue, and hardness. There were studies of crushing, punching-out pieces, or fracturing human bone. The living bone was compared to dead bone, dried bone, bone fixated with alcohol, and embalmed bone (Evans, 1973).

(4) Dummies

The use of surrogate manikins (dummies) has become widespread in crash testing. It started with relatively unsophisticated hard plastic, partially instrumented, average-size, male dummies. Gradually, after much research coordination, more life-like (biofidelity) manikins emerged. Although they still have significant shortcomings, there is a considerable database for purposes of data comparison and the generation of useful biokinetic information. Repeatable tests can be predicted on standard manikins that have industry-wide acceptance.

> The ultimate test of any safety design change is its influence in the real world of accidents. A given development may look exceedingly beneficial in the laboratory

or on the test track, but until its effect on a known population of collisions is established, it is impossible to quantify the benefits and therefore justify its more widespread introduction.

(MacKay, 1973)

(b) Crashworthiness

(1) Deceleration curves

When a vehicle is involved in a collision, there is a crushing of metal. The resistance to that crushing reduces the speed of the vehicle until the vehicle is stopped or the vehicle disengages from the object that is struck. The soft crush of sheet metal produces very little slowing and creates low deceleration pulses in the vehicle. A hard metal encounter, such as the structural members of the frame rails, produces more resistance, more deceleration, and more energy transfer. When the engine is engaged, there is virtually no crush and it produces a high amplitude spike of deceleration transferred to the vehicle. The series of decelerations, over the hundred or more milliseconds of a crash, may be called a crash pulse or deceleration curve.

The series of deceleration pulses first affect the vehicle, then the interior components, and then the occupants. The more severe and abrupt the deceleration, the more severe the possible injuries. Each company may have maximum tolerable deceleration spikes or a maximum average pulse for a set time interval pulse duration. The objective, obviously, is to keep decelerations low in order to minimize occupant injuries or to identify and correct or compensate for unwanted types of deceleration that may be delivered to the occupants. The collision sequence involves *relative* movements between the occupants and the vehicle interior as reflected in the deceleration curves. Biokinetics are a product of the timing of decelerative events and the magnitude of the forces that might be experienced. The essence of crashworthiness is energy management and the control of crash deceleration pulses. Examples of deceleration curves are shown in Figure 10.1.

(2) The square wave

The square-wave deceleration curve is an important and useful theoretical construct. It infers or describes a smooth uniform deceleration during a collision. The design objective is to achieve a controlled constant crush, without imparting sudden changes in velocity to the vehicle and the occupants. The readings taken by accelerometers actually provide a somewhat irregular or jerky readout, not a smooth curve. The readout varies as to the placement of the accelerometers; for example, on the vehicle structure, on the seat, or on the occupant (dummy).

The square wave must have a beginning and ending, i.e. changes in deceleration. If those changes are smoothed-out, the curve might look *triangular* in shape, the first slope being increasing deceleration and the second slope showing decreasing deceleration. This concept evolved into a *bell-shaped* (normal) curve as a design objective. There would be an onset of deceleration, a plateau, and then a decay.

There are those who are proponents of an *early peak* curve in which significant deceleration occurs early and there is a long slope downward when biokinetic events are occurring in the interior of the vehicle. This promotes the key concept of a safe *ride-down* where the restrained occupant is firmly connected to the vehicle and moves with

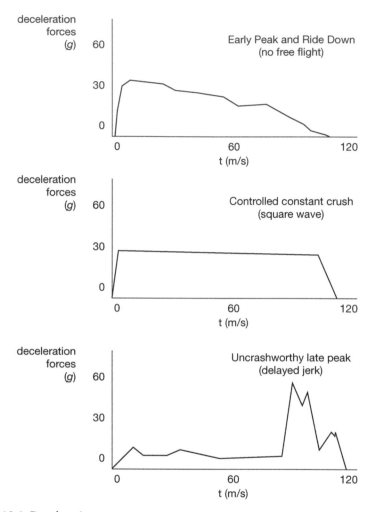

Figure 10.1 Deceleration curves

the vehicle during the collision sequence. In other words, ride-down suggests no interior contact by the occupant unless there is some intrusion into the occupant's cocoon or safety cage.

The terms commonly used include velocity, the rate of change in velocity (acceleration, deceleration, or *g* are used interchangeably), and the rate of change in acceleration (measured as peak gravitational units, *g* per second, or *g*/s) (Stapp, 1986). The rate of onset of acceleration (jerk or pulse rise time) is an important variable. Extremely high rates of onset are tolerable if they are of short duration (Viano and Gadd, 1975; Gurdjian, 1970).

(3) Injury tolerance

As a general concept, a fully restrained human occupant should be able to *survive* a 50 mph frontal collision with an unyielding barrier (McFarland and Moore, 1970, p. 146). This still remains a design ideal or safety design objective. Survivability, if there is a head impact, depends on the *g*-forces and their duration. The commonly used guidance diagram for the force–time relationship is the so-called Wayne State Curve (Patrick, 1965), which is for forehead impacts on a hard flat surface. A variation of that curve is shown in Figure 10.2, and there are many variations depending on the force vector and location of the skull.

The crash test results should be assessed against the desired deceleration curve and the head impact tolerance curve. However, it is very important to take the next step, which is to determine the occupant's biokinetic responses and interactions during the crash pulses shown on the deceleration curve. In essence, the issue is: what is the effect on the occupant, for better or worse, of the forces represented by the decelerative curve?

(4) Control of deceleration

The primary objective in crashworthiness is effective energy management, and a number of methods can be used to achieve that objective. The most familiar is to design a

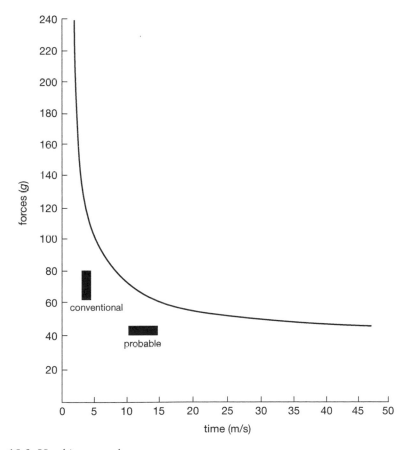

Figure 10.2 Head impact tolerance

crushable front end structure that will result in a progressive and uniform vehicle deceleration. This is without major spikes, irregularities, skewing, or jerks in the deceleration curve. The front end may be divided into four or five equal-distance crush zones, each having about the same crush resistance or deceleration value. This may involve relocating some components to other zones. It may involve adding crushable structures or crossmembers. The engine mass may be made displaceable in some manner.

The forces generated during crushing should be distributed, dissipated, and diverted from the occupants. For example, a section of high-strength steel might be placed in the front rails for added resistance and both rails tied together so deformation occurs in both, even in an offset collision. Fairly complex one-piece front rails have been achieved with hydroforming.

The blows might be softened, cushioned, or smoothed out by seat cushions, seat isolation systems, and structural members that distribute and modify the loads.

The engine hood may be notched so that it folds rather than being forced through the windshield. Any intrusion into the occupant compartment is to be avoided.

On the interior, airbags and soft or crushable materials can be used on the instrument panel, headliner, and other interior surfaces.

Each component should be evaluated in terms of compression, tension, and bending stiffness within the context of the vehicle's overall crush characteristics (i.e. a systems analysis). The initial penetration and component interaction (compatibility) are important and intersections (separation at interfaces) present the greatest problems.

The location of impact (crash modes) may include full frontal and offset frontal into rigid or deformable barriers. Some may be angular impacts. They may be at something over 30 mph, such as 31 mph (50 km), or something over 35 mph (56 km). The intent is to determine whether effective control of deceleration is achieved at varying speeds and impact vectors.

Some engineers use the deformation at the B-pillar and rocker as a visible and measurable criterion and others use crush intrusion into the occupant (passenger) compartment as a visual indication of the integrity of the protective cocoon.

For multiple vehicle collisions, the height of the main structural elements may be a mismatch when the target vehicle is a passenger car and the bullet vehicle is an SUV.

(5) Pole testing

Vehicle impact with a roadside tree, telephone pole, light standard, or highway sign support may be rather specific in terms of vehicle crush. All the impact forces may be concentrated in a small area of the center of the front-end structure. There are vehicles on the road that have very little crushable structure between the front rails, so an impacting pole has little resistance until it strikes the engine block. This results in a delayed, high magnitude, crash pulse (jolt) in the deceleration curve.

If there is a low front-end crossmember between the rails, it may tend to pull the frame rails toward the bending crossmember. This additional frame bending serves to absorb more energy and prolongs the crash pulse. However, near the end of the vehicle crushing, there may be a pitch-up or lift-up of the rear of the vehicle. This is because the crossmember is located below the forward-moving center of mass of the vehicle.

The better design approach is to have two crossmembers, one at the end of the frame rail (the lower crossmember) and one above the radiator, near an extension of the beltline, or close to an extension of the cowl line (the upper crossmember). In other words, a second crossmember in the front structure is above the vehicle center of mass. This should reduce pitch-up or lift-up, reducing vertical forces on the occupants. It also provides for a more crushable box-like structure to smooth out the deceleration curve and to help distribute the impact energy better.

Pole testing has been somewhat neglected, perhaps because a concentrated force causes deeper penetration of the vehicle. An offset impact is somewhat like a pole impact, as compared to a full frontal barrier collision. It also may result in some vehicle rotation on rebound. However, pole testing is important because of its effects on the occupant biokinetics.

(6) Rear testing

Rear-end crashes are common. The vehicle crush analysis is similar to that for frontal collisions. The rear structure is easier to design for uniform crushing or a relatively smooth deceleration curve, because it may be a box-like welded structure. It should be designed to protect the fuel tank. The hard objects may be the rear axle and the two wheels. However, there are many variations in rear-end design. Some rear-seated occupants sit perilously close to the rear bumper, with little crushable material to protect them. In a rear crash, the occupants move toward the incoming bullet vehicle and need proper head restraints.

Welding is a large part of the basic manufacturing process for vehicle assembly. There have been instances of misaligned body panels, inadequate or missing welds, or other problems affecting the integrity of the body shell. Poor welding can drastically affect crush in the target vehicle, permitting major penetration by the bullet vehicle, and significantly reducing calculated impact speeds. In one case, the actual speed of impact was less than half that originally estimated.

(7) Side impact testing

There is little crushable material in the sides of most vehicles, so a different design approach may be needed. On the interior, some cushioning may be provided by side airbags (shoulder and head). The buttocks may be restrained by deeply recessed seat cushions. There may be side wings that are automatically actuated to hold the lower torso. This assumes an effective occupant restraint (belt) system.

The impact forces may be transferred from the A-pillar, B-pillar, and C-pillar to specially designed upper and lower crossmembers, around the occupants, to the other side of the vehicle. A rigid safety cage provides protection from intrusion. The critical area is at the rocker and lower floor pan, so this perimeter location may need strengthening. A beam-like structure may result in a step-down entrance for passengers. There may be a regulatory requirement to provide a safety brace within the door. The door should be able to distribute the impact forces to the B-pillar and other structures.

The door and its frame have become increasingly important as a prime structural member to resist side, front, and rollover forces. There are more sliding doors, forward opening doors with new locking devices, and oversized doors. Because of the importance of door systems, side impact testing of the entire vehicle is necessary. The doors are the repository of many control devices, such as window lifting mechanisms, electronic door locks, rear deck releases, fuel tank filler releases, powered mirror adjustment controls, as well as compartments for small objects and toll coins. The simple door is becoming a vehicle subsystem.

(8) Rollover testing

Rollovers do occur. Early testing involved cliffside testing in which the vehicles were rolled over and they tumbled down a cliff. There were early standards on ramp testing that provided a more standard rollover situation. Some vehicles were rolled sideways off the beds of moving transporters to achieve three or four rollovers. Roof crush and intrusion of the passenger compartment were the critical variables. But serious questions arose as to the repeatability of the tests. In other words, seemingly identical test conditions produced different rollover situations and results. This does not diminish the need for rollover testing, particularly with the increase in high center of gravity vehicles, those capable of being loaded to higher heights, and those with large bodies on a conventional chassis.

(9) Other vehicle tests

It may be difficult to apply standard crash test protocols, procedures, and criteria to a wide variety of automotive vehicles. But tests with important design safety implications should not be neglected, even when the intended function seems to be the only priority other than production schedules.

Agricultural vehicles

Agricultural vehicles operate on farmland but often use public roadways. Crash-worthiness should be a consideration, but more important are static and dynamic stability maneuverability under loaded and unloaded conditions, articulation and pivoting of the vehicle or prime mover, and function-specific safety problems such as material conveyor bands, screws, rollers, stops, and alarms.

> In agriculture, it is found that over half the overturning accidents with machines are due to exceeding their slope capabilities, while only one quarter can be blamed on carelessness or misjudgment of the driver. The purpose of research into stabillity on slopes is to provide safety information which the driver currently lacks.
>
> (Hunter, 1993, p. 337)

Military vehicles

Military vehicles range from heavy tanks to light reconnaissance vehicles, from troop carriers to unmanned robotic devices, and from rough-terrain vehicles to those very similar to civilian vehicles. The image of military vehicles is spartan, functional, durable, and single-mission-oriented. Armored vehicles may have a somewhat low presented surface profile, with interior weapons loading and fire control, communications and navigation electronics, fuel and engine, bulletproof glass and periscopes, and the steering and other controls. The human occupants may have to adapt to the leftover interior space. If the driver is in a nearly supine (on his back) position, what safety problems might be created? The tank may have a fairly rigid and durable suspension system, but what whole body vibrations and uncushioned shocks are delivered transversely to the human body? What stresses are involved for a supine driver, under battle conditions, with vehicle interior weapons loading and firing, temperatures and humidity extremes, and the normal fear of being hit by enemy artillery fire, anti-tank hand-held weapons, heat seeking missiles, or thermite grenades and bottles of gasoline? Under non-battle conditions, what are the short-term and long-term effects of cramped quarters, limited visibility, sleep deprivation, fatigue, fording streams, navigation uncertainties, team peer pressures, wearout and parts replacement, and the unknown conditions they may confront?

There have been many studies of the human factors involved in the operation of military vehicles, but much more needs to be done because of changing demands on human performance, more sophisticated and complex vehicles, and differing command structure and policy expectations. While crash testing may seem a remote requirement, the specialized information could be transferred and utilized in the design process. Many vehicles should be subject to standard crash testing. Within all the constraints of military procurement, occupant safety should be considered early in the design process, not left to a final design review for compromise or resolution.

Construction equipment

Construction equipment has similar problems in terms of variety and dedication to a specific function. Some equipment is included in the normal public roadway traffic mix, some has a priority on the highway during construction activities, and some is a sitting target on the sides of high-speed roadways. Some could benefit by crash testing, others would not. All could benefit from the application of safety principles and data derived from the crash testing of other vehicles, and the information revealed by more formal and detailed accident reconstruction.

(c) Compliance testing

Some crash testing is a mandatory government requirement to physically demonstrate compliance with domestic vehicle safety standards. Unfortunately, passing such tests became a goal, something to be exceeded by as little as possible, for various reasons. Barrier testing at 30 mph was not considered a bare minimum or a starting point years ago. The concept of a world-class car that could easily exceed all international standards was not accepted. There were many rationalizations starting with costs, involving technical feasibility, and ending with a defensive posture against any government regulation as infringing on a competitive free market.

Compliance testing has been performed by manufacturers, independent laboratories, universities, insurance-funded entities, and private parties. The non-government test facilities gradually increased the testing, both in a higher velocity of impact and where the impact was delivered. Private parties and professional seminars performed multi-vehicle collisions. Test facilities supported by the insurance industry developed rating systems and the results appeared in vehicle advertisements and served as incentives for vehicle brand managers. Test facilities operated by universities provided considerable data and interpretive comment.

It was found that mandatory compliance testing intended for one purpose might reveal other safety information. For example, some fuel-tank integrity testing revealed severe manikin head movements and impacts, upper torso ramping, seatback problems, and the effect of head restraints. It should be clearly understood that success on some tests may not predict success on other tests.

(d) Component testing

Most vehicle manufacturers have fairly detailed design specifications for vehicle components. They have test requirements to determine whether the components meet or exceed the specifications. All too frequently, the test requirements may not be entirely adequate or relevant to the real needs. Sometimes they are waived in one form or another to meet preproduction schedules. These situations may not create or manifest problems until well into a production run when corrective action may be too costly to institute. This suggests that compliance testing should be audited within the company.

There is a trend to delegate design and test responsibilities to first-tier suppliers. The vehicle assembler may have only generalized design specifications. These are intended to permit creative solutions at lower cost. The supplier formulates the detailed design specifications, the test requirements, and the quality control obligations. The psychological distance between assembler and supplier might be great. The consequences of failure to meet the test specifications may induce the creation of innovative approaches to test compliance.

Component testing is not equivalent to system testing, testing the whole vehicle, or testing the driver in the vehicle. It is a necessary first step to weed out defects, deficiencies, and functional inadequacies early on to keep costs low. It rarely duplicates the effects of whole-vehicle crash testing.

(e) Competitive race testing

Early competitive racing in Europe provided excitement for the public and publicity to establish brands. The objective was to help sell automobiles. Cross-country races also generated publicity and sales, but also served to identify problems associated with durability, speed, and steering responses under adverse road conditions. In the United States, racetrack vehicles bore little resemblance to production vehicles. Even stock-car racing was with modified vehicles. However, such race events did represent a type of punishing ordeal for the driver and the vehicle. It was a form of accelerated life testing and failure-inducing process that could reveal weaknesses that could be corrected.

Racetrack drivers are not representative of other drivers and they are paid to be daredevils. They experience fairly frequent, high-speed, multiple-impact, and recorded collisions that have proved to be valuable sources of information. Instrumented drivers and vehicles have become more commonplace. They often demonstrate that the human body can withstand vehicle impacts at very high speed, if properly restrained in a crushable structure. Some unexpected injuries and deaths were also instructive, although undesired. They have demonstrated that tires and wheels that come loose in a collision can be very deadly weapons to other vehicles and those in an audience located at a seemingly safety distance behind a barrier.

One of the more important lessons learned in automobile racing was to put the important driving information in the driver's line of sight. This prompted the development and use of head-up displays (image projections on the windshield) as a driver aid. The search for simpler and less costly techniques has included placing digital displays in clear plastic on the top of the steering wheel and the use of optional audible messages to the driver.

(f) Proving-ground testing

Generally, the first intensive and more realistic whole-vehicle testing is performed at a company test track or proving ground. This is an important event in the vehicle development cycle. Additional testing may be performed at high-temperature arid desert facilities, low-temperature arctic facilities, cross-country locations or simulated adverse roadway conditions, and by fleet testing on the public roadway. It is possible to accumulate high mileage fairly quickly at high speeds, in a form of accelerated life testing. There are many tests that can be conducted with humans as drivers.

Object avoidance tests may involve simulated deer, dogs, or other animals. They are made simply to dart out in front of a vehicle and the driver must take evasive action. Such tests are more common in Europe. They measure vehicle stability, control, and rollover potential.

Collision avoidance tests may involve driver response to warning signals or the actions of other vehicles. This might include test vehicles with human eye fixation recorders (object, distance, and time) to objectively record data otherwise unavailable. It may be to extend, more realistically, the findings on mechanical simulators (training and research devices) that may determine task loading, time to locate and actuate controls, read and interpret displays, and to understand input data from telematics devices.

(g) In-field testing

Some production vehicles may be allocated to employees or focus groups to identify what has pleased them, displeased them, or caused them some trouble. Their comments may be very instructive.

Most in-field or post-sale testing is observational or data collection efforts pertaining to actual on-the-road consumer usage, e.g. observation of cell-phone use at particular locations, times, weather conditions, and in various vehicle types. Questions may be asked about navigators, vehicle options, and driving habits.

Headlight illumination patterns, distances, aiming, and automatic dimmers can be evaluated at the component testing level and during proving ground testing. But in-field testing and observation are important. For example, a high-intensity headlight might pass aiming tests while the vehicle is at rest, but when it traverses an irregularly paved or bumpy roadway the oncoming driver may be blinded by pulses of glare or the glare may be continuous if the vehicles were not properly made ready for the purchaser (perhaps only one headlight was adjusted for proper aim).

Certainly, in-field testing may be the only means to determine the effectiveness of anti-theft devices. In-field testing provides information on the unexpected events that may be quite foreseeable to those outside the industry. It may reveal custom and habit of various demographic or cultural subgroups. How else can the driver's behavior be studied at highway–railroad crossings?

In-field testing data may be collected from road-test publications, dealer reports, sales and marketing data, repair shop parts orders, direct customer satisfaction surveys, and salvage yard observations.

Of course, this form of customer testing provides the most important source of information: the accident reconstruction data resulting from real-life collisions.

11 Accident reconstruction

(a) Introduction and objectives

Appropriate accident reconstruction is an essential task, assuming that causes should be identified so that preventive remedies can be determined. In other words, the ultimate objective is product or process improvement for future risk reduction. There may be some immediate subordinate objectives, such as fault determination for police action, for criminal or civil liability purposes, for insurance reports, for temporary site alterations or immediate equipment modifications, and for recall or regulatory actions.

Regardless of the purpose of an accident reconstruction, only time-proven, peer-acceptable, fact-based techniques should be utilized. A commonly accepted foundation for the analysis increases credibility and acceptability, while reducing conflicts in the findings and conclusions of others who might perform a similar reconstruction. The methods are generic, whether applied to passenger vehicle collisions, single earthmoving vehicle turnovers, an aircraft crash, some recreational vehicle impacts, a machine tool industrial accident, or any other accident situation. The content and words may be different, but the so-called 'laws of physics' remain the same. The time-honored methodology is immediately productive and it provides comparability in engineering terminology or replication in scientific research terms.

The accident reconstruction process generally begins with a site and vehicle investigation, proceeds to a literature search and the marshaling of supporting documentation or foundational information, and, finally, the preparation of reports and graphics. A follow-on effort is often advisable. While the process seems rather straightforward, there are always variations between investigators and analysts that could result in significant differences, complications, and even advocacy for vested interests. A bias might result from the use of more sophisticated available technology, the level of detail utilized, or philosophical differences. For example, in terms of philosophy, some safety specialists advocate faceless anonymity, faultless reports, or no-repercussion promises, since they believe that more honest reporting will occur if there is no possibility of suffering personal fault accusations or possible company liability. That process generally involves self-investigation and admissions. This chapter will emphasize the independent investigation, the broad scope required for an adequate accident reconstruction, and techniques to avoid common mistakes and omissions.

(b) The initial investigation

(1) *Photographic documentation*

As an accident scene is approached, photographs should be taken to record the scene before any further inadvertent damage could occur. Just the movement of people at the scene, including the investigators and rescue personnel, could move fragments of the vehicle on the roadway, obscure tire marks on a dirt shoulder, or bend sheet metal as they lean on the vehicle to look for something else. The photographs should be taken from all major directions to record the general scene. This includes all sides of the vehicles, the approach path of each vehicle or moving object, and photographs taken to document the beliefs and opinions expressed by witnesses. Whether an intersection collision, a construction equipment accident, or an industrial injury, an inquiry should be made of any changes or modifications made after the collision, accident, or injury. Such changes should be photographed to compare with any unchanged exemplars.

There are two important *caveats*: first, take at least two photographs of every view, preferably with two different cameras (a primary and a backup camera); and second, photograph everything, since something overlooked may become important later and human perception is such that key evidence may not be seen or properly interpreted until a review of the photographs occurs.

It may be important to use a camera monopod or tripod to maintain a constant camera height and location. Sight obstruction or blind-zone photographs should be taken at the eye height and the location of the person so encumbered. If police or company photographs are available and changes noted, similar photographs should be taken to record the before-and-after modifications.

For intersection collisions, all signage and other traffic control devices should be photographed. Orange cones may be used as locators or distance markers. The second round of photographs may include measurement rulers to demonstrate the accuracy and method of determining distances and dimensions. The location and direction of each photographic view should be marked on a rough sketch of the accident scene. Each photograph should be entered in a log with a notation of where, how, when, why, and by whom it was taken.

Cameras taking still photographs with film are preferable because they are difficult to alter. Digital cameras are more versatile, with high zoom potential, quick enlargement and printing, and large shot capacity, but are easily altered on some desktop computers. Videotape may be desirable to demonstrate key behavior, equipment movement, or changing views from a moving vehicle. They may assist in the preparation of computer-generated video graphics or large exhibits that can have accident overlays (transparencies) placed upon them.

Aerial photographs may provide important details, otherwise unseen, of a collision site. Various road gouges, tire marks, and debris trails may become visible. Cameras on balloons, fishing poles, and work platforms of boom cranes have been used for overhead shots of vehicles in a collision.

An automobile component replacement manufacturer experienced several serious injuries in its metalforming and joining area. Many of the punch presses, press brakes, and bending machines were poorly guarded. One old press had continual cycling until turned off by a wall switch and this was considered illegal by a retained independent safety consultant. Some safety recommendations were made, directly to the chief executive officer, as to emergency guarding. But photographs of each machine were enlarged by office copiers to permit illustrative sketches of the locations on each machine where safeguards and warnings should be installed. Illustrative copies of temporary warnings and the sources of permanent warnings accompanied the photographs. The metalforming equipment of this aftermarket supplier was almost entirely previously owned (used), obtained at auctions. Such used equipment produced products with major dimensional variations, so the allowable discrepancy rate was high. These variations resulted in defects as originally manufactured and during wearout. The defects could produce an ignition source during vehicle collisions. The presence of an ignition hazard was unrecognized by the manufacturer or by the distributors and retail sales outlets that were supplied with the product.

(2) Measurements and diagrams

An on-site investigation provides the opportunity to observe, measure, and record a variety of relevant details concerning the presence of the roadway, its coefficient of friction, the road slope, visual obstructions, the sight distances, the traffic flow, the traffic control devices, possible skid marks, the accident vehicles, their positions at rest, the deformations on the vehicles, and many other factors or considerations. Each accident shares some commonality with others, but each is uniquely different. A preliminary estimation can be made of the approach pathways of vehicles involved in a collision, the impact location, and their post-impact trajectories. These are in the form of hypotheses against which all the evidence can be systematically compared and evaluated. The hypotheses may be modified to fit the facts and then serve as a model to be verified or rejected, in part or in whole, by a subsequent in-office analytic reconstruction of the accident.

There may be witness marks (scrapes and gouges) on the road surface that may suggest a post-impact movement of a vehicle. There may be tire skid marks that can be used to estimate the pre-impact speed of the vehicle. There may be a spiderweb type of windshield glass fracture, the center of which may have been caused by a human head impact. The slope or grade of the roadway and its coefficient of friction could be measured. Visual obstructions and their resultant blind zones could be recorded. All of this data may result in rough sketches or diagrams, with roadway and curb outlines, notes as to signs and signals, and, most importantly, reference points, landmarks, and baselines should be included. A table of measurements, between points on the street or in the vehicle, should be prepared. This assists in preparing accident situation maps and computer simulation graphics. This preliminary site evaluation may be redrawn for greater clarity and to scale at a later date, but the original sketches and notes should be retained in the accident file.

Some investigators distinguish between steering yaw or side sliding marks, acceleration scuffs or melted rubber tracks, and collision-induced scrubs or brushmarks. There may be an estimate of take-off, airborne, and landing speeds for vehicles running off cliffs, embankments, landings, or bridges. The investigator may utilize the injury patterns to determine injury causation. For example, a passenger in an automobile was forced downward by an upper torso belt routed over his shoulder from a near-floor mount. It appeared to be a cervical injury because the collision was at a relatively low speed and his shoulder was depressed by the belt. A tear-apart or degloving of the seat back revealed a major deformation of the steel back support. Major downward forces had been applied to the passenger. A subsequent MRI image revealed a major disk herniation and extrusion in the pelvic (lumbar) area. *Caveat*: Occupant injuries should be considered as an integral component in the physical and human chain-of-causation that characterizes modern accident investigation and reconstruction.

Informal conversations with nearby residents may indicate that the collision site has a history of speeding vehicles, a high accident rate, a pedestrian injury problem at certain time periods in the day, a near-miss history of turning incidents, or changes in the duration or character of the traffic signals. Information about motor vehicle ownership may lead to past owner or operator interviews to determine vehicle repair, modification, recall, or damage incidents.

If there were a vehicle fire, the point of origin should be determined. It might be an engine compartment fire, a passenger compartment fire, or a fuel tank that was crushed in a rear-end collision. What was the burn pattern? Was there evidence of arson? Were combustible materials being carried? In one example, a severely burned driver had filled a gasoline can and carried it in the occupant compartment to assure that it would not tip over or leak. He did not know that the gasoline container was of a type disapproved of in some countries because it did leak and was not fireworthy. The burn pattern of another vehicle indicated an origin (lowest point of a v-shaped burn pattern, narrowest flame spread, and greatest heat damage) at the right rear wheel well. There was no evidence of drop-down burning such as gasoline on the ground. A badly deteriorated exhaust pipe was found that was located very close to the tire and the vehicle had been allowed to warm up in a stationary position, without the benefit of the air flow of a moving vehicle to reduce high temperature buildup.

In still another case, a gasoline tanker exploded while being filled with fuel. Static electricity was the likely ignition cause and gasoline vapors the fuel source. To assure proper grounding of the vehicle in the future, a grounding resistance detection system was installed at the gasoline terminal (to detect the electrical potential difference between the gasoline tanker vehicle and the ground of the terminal).

Measurement of vehicle crush is extremely important. Various devices are used to establish a perimeter base reference line. Then, measurements are taken between the reference perimeter and the crushed or distorted area of the vehicle. This is done in a series of crush zones, so there may be six crush zones with equally spaced measurements of depth of crush or vehicle bowing. The measurements should be taken perpendicular to the crushed surface, at the height of the deepest penetration. Based on the stiffness of the vehicle, the extent of vehicle crush may reveal the velocity (speed) of impact, the magnitude of forces applied, and the general crashworthiness of the vehicle. There are electronic sighting systems, used to take field measurements from a reference point to selected points on the vehicle (usually to a temporary marker or numbered patch). This data may be compared to the same points on the original vehicle, or it may be used to produce automatically a scale drawing at the accident site.

The skid marks left on the roadway may indicate that one vehicle was traveling 50 mph (80 km) in a 25 mph (40 km) posted area. The deformation on the front of V-1 (vehicle number one), and matching deformation on the driver's side of the vehicle, may suggest that V-1 impacted the side of V-2. A passenger that is found on the roadway suggests occupant ejection of a non-restrained (unbelted) person. An opened alcoholic beverage container or package of marijuana in the vehicle interior suggests that further investigation is needed of possible driver intoxication and impairment. However, these are preliminary inferences that need to be verified by a more detailed analysis, consideration of the totality of the accident, and further study of the relevant available literature and other information collected for the analyses.

There may be a need for a second or third visit to inspect a vehicle, either to recheck some neglected item that now seems important or to bring a specialist to examine some unique feature or condition. For example, samples of blood stains and flesh might be taken, teeth marks examined in greater detail, light bulb filaments that are whole or broken photographed more closely, and a more specific comparison made of the paint transfer from one vehicle to a vehicle that had been sideswiped. There may not have been locked wheel skid marks on the roadway, but the tire treads could reveal a particular pattern characteristic of the full engagement of antilock brakes, and all of the tire marks may need a further differentiation and explanation. The seatbelts may exhibit marks that show that an occupant may or may not have exerted considerable force against a locked seatbelt, but the manufacturer's identification is necessary to determine whether they are original equipment, an aftermarket replacement, or have material degradation in belts supplied by one vendor. A missing element in a preliminary speed and control calculation might be revealed in a second inspection to be a smooth bleeding asphalt area in a vehicle's initial approach pathway to a collision. Supplemental information on cell-phone usage, occupant bruises, and the possible movement of objects within the vehicle interior may warrant a reexamination of the vehicle and the accident site. The objects (such as a tire) may have been removed from the accident scene and not been known about at the time of the first inspection.

In some situations, only a quick bare-bones investigation is necessary. For example, two vehicles were stopped at a marked crosswalk to permit a young child to cross the road. An ambulance approached the rear of one vehicle, with flashing lights, and sounded its horn so that the automobile would clear its path. The vehicle moved out of the way, fearing an emergency, and was broadsided by another vehicle in the intersection. Was the ambulance (V-1) at fault, was it the vehicle (V-2) that moved out of the way, or was it the vehicle (V-3) that collided with V-2? The question was formulated by an insurance adjuster and referred to a law firm to determine fault. It was, in essence, a legal question requiring the application of local legal principles, rather than a technical question requiring sophisticated accident investigation and reconstruction. Only the police accident report and a few witness statements provided the limited factual information deemed necessary to resolve the legal issues. In other words, the character and extent of the accident investigation is needs-driven, so the specific purpose defines and limits the activity.

The basic purpose of the on-site investigation is to observe and identify all relevant information, measure and make diagrams of all important considerations, and to evaluate and document all factors that should be considered in a subsequent accident reconstruction.

(3) Traffic collision (police) reports

The police report of an accident is important because it documents what occurred shortly after the accident, before the evidence is taken away and the scene changes. Such a report is given great weight to the extent that it describes physical facts (evidence), documents measurements taken by the officer, and records the scene by means of photographs. Lesser weight is given to diagrams because they may be conclusionary guesses and based on the subjective opinions of those who claim to be eye-witnesses. The names, addresses, vehicle identification, and accident location are generally considered reliable. The witness statements are usually taken at face value until controverted. They often indicate areas of conflicting opinions or possible disputes that need special attention. Since much of the report is 'as told to him' and is a search for immediate causation, such as 'excessive speed for conditions', in order to complete a checklist on the report, the final conclusions are generally less credible than the factual portion of the report. In general, the conclusions, opinions, and impressions are given far less weight and are generally inadmissible in court proceedings. The weight given often depends on the training and qualifications of the officer, the technical or legal complexity of the issues, and the character and extent of any subsequent supplemental report that can corroborate the findings. The police reports are generally helpful as to the presence of skid marks or tire marks, the time of day or night, the weather, lighting, roadway surface conditions (wet, ice, snow), sobriety, and the presence of traffic-control devices.

(c) The search-and-marshal effort

Immediately following the initial investigation, a search should be made for all sources of information that might be helpful in evaluating the facts and the hypotheses generated in the investigation. Subsequently, material deemed relevant should be marshaled in one file so that a solid foundation is available for analyses, the reconstruction conclusions, and the opinions that might be offered in adversarial situations.

The search may range from a quick review of the books, journal articles, and various reports immediately available to a detailed patent search (the patent file and wrapper). There may be valuable information in owner's manuals, service manuals, vehicle specifications, and a variety of publications describing vehicle features and performance. There may be stopping distance data (from 60 mph to 0 mph) and vehicle crash stiffness coefficients (frontal, offset frontal, and side stiffness). There are portable braking test devices that measure acceleration and deceleration, speed and time, distance and grade, drag factors, and g forces (lateral, average, and peak).

There are firms that can provide complete reports on over 2000 NHTSA crash tests and those reports include crush measurements, photographs, dummy responses, and accelerometer graphs. For example, for one vehicle in a frontal collision with a fixed rigid barrier, at an impact speed of 35.1 mph (56.5 km/h), the average crush distance was found to be 17.7 inches (44.9 cm), and measurements at the center of six different zones ranged from 12.6 inches to 20.2 inches.

There are automotive publications describing the reasons for changes in company and government requirements. There are lists of recalls, federal preliminary evaluations and engineering analyses, and liability lawsuit allegations. There are private testing grounds and private sources of data. For example, one firm provides passenger vehicle and motorcycle specifications including weight, length, width, height, wheelbase, wheeltrack, front overhang, hood height, bumper height, turn circle, wheel radius, ground-to-hood, bumper-to-hood, bumper-to-windshield distances, etc. Another firm provides headlamp data, acceleration speeds and distances, braking distances, interior dimensions, materials used, and center-of-gravity information. Another provides information on crash tests of vehicles into guardrail terminals, into pedestrians, into bicyclists, and into signposts. Others provide a car-to-bus impact video, damage repair estimates, sun location, twilight times, weather condition, high accident intersections, and last bar visited. Almost anything needed is supplied by the accident reconstruction community.

It is important to view crash videotapes several times since important information may be revealed on a topic not suggested by the test title. For example, tests to determine possible leakage from gasoline tanks during rear-end collisions also show the violent head and upper torso movements of the test manikins and the type of airbag deployment.

(d) Analysis and reconstruction

(1) Vehicle crush

One reconstructionist, using a simple computerized *crush model* (EDCRASH, see p. 152), determined that a vehicle impact speed was 19 mph (30.6 km/h). This is a damage or crush volume model (six crush depths and a stiffness value), derivative of other CRASH (see p. 151) models, that provides a barrier equivalent speed. By using a simple rule of thumb, that each one inch (2.54 cm) of vehicle crush equals 1 mph (1.6 km/h), a 27-inch (68.6 cm) (maximum) crush distance would result in an impact speed of 27 mph (43.4 km/h), for localized, focused, or pole impacts. Using a police report that indicated a maximum crush of 24 inches (60.9 cm), the speed would be 24 mph (38.6 km/h), based on the rule of thumb method. Still another formula indicated an impact speed of 25 mph (40.2 km). The delta-V speeds were 23 mph (37.0 km/h) (EDCRASH), and 33 mph (53.1 km/h) (formula). The accident reconstructionist concluded that the impact speed was probably in the 'mid-20s' (38 to 42 km/h). The rebound or restitution was about 20% for an in-line frontal pole impact.

The general rule is the greater the crush and the greater the vehicle stiffness, the higher the impact speed. But the crush characteristics of each vehicle differ, depending on such factors as the configuration of various structural members in the crush zones. In the preceding evaluation, the impact speeds varied from 19 to 27 mph (30.6 to 43.4 km/h) and the delta-V speeds from 23 to 33 mph (37.0 to 53.1 km/h). These major differences are to be expected because of the different reconstruction models and the quality of the data being used.

Another accident reconstructionist, dealing with the same vehicle and accident, for a pole-impact maximum crush of 21 inches (53.3 cm), using a formula based on the impact data, estimated a speed of 20 mph (32.2 km/h). This was a delta-V of 20 mph (32.2 km/h), a barrier equivalent speed of 9 mph (14.5 km/h), and a peak *g*-loading of 16. By another method of calculation, still using crash-test data, the impact speed was 23 mph (37 km). Still another estimate was based on crash tests of similar vehicles, using a k-2 stiffness (11.3) factor, and resulted in a 19.6 mph (31.5 km/h) impact speed. His best overall estimate was a 20 mph impact (32.2 km/h). Allowing a curb impact reduction of speed of 5 mph (8.0 km/h), the original speed prior to curb and pole impact was about 25 mph (40.2 km/h).

His range of impact speeds was about 20 to 23 mph (32.2 to 37.0 km/h), as compared to the first reconstructionist's 19 to 27 mph (30.6 to 43.7 km/h). His best estimate of 20 mph (32.2 km) (apparently biased toward the lower value) was compared to the mid-20s estimate (biased toward the upper values). Of significance was that all estimates were below the 30 mph (48.3 km/h) value used in early crash tests to determine regulatory compliance for various safety features.

(2) *The crash-event sequence*

A single-vehicle crash event sequence generally occurs over a time period of 100 to 140 milliseconds, but often in less than one-tenth of a second. This event sequence starts with an initial contact (impact) with another object and continues to disengagement, restitution, rebound, or additional vehicle movement in an altered direction. Each event sequence in multiple collisions is analyzed separately.

The accident reconstructionist may prepare a timeline diagram to show when each event occurs and the interactions that result between the occupants and the vehicle interior, components, and devices. The timeline includes the force-crash pulse or decelerative changes (g-loadings) that are experienced during the collision. The crush of soft metal produces low decelerative changes and the crush of more substantial materials produces higher deceleration or resistance to crushing and bending. The crash pulse may be relatively smooth if there are fairly uniform crush zones (controlled crush by design), but may reveal some high-magnitude pulses, spikes, or reactions at various points on the timeline. There are various company and published criteria as to such force–time pulses, including maximum vehicle deceleration criteria, average vehicle deceleration criteria, and occupant deceleration (biokinetics) criteria for head and upper torso that may or may not be exceeded. It is the crash pulse that is very important in reconstructing the most significant events that occur during a collision. For example, in a typical front-end collision, the upper torso restraints (shoulder belts) would have their slack removed (from looseness or comfort) as the occupant moves forward, since the occupant continues to move forward until restrained in some manner as the vehicle is decelerated. The belt spools out of its retractor until the belt webbing locks by an inertial or pendulum locking mechanism. A pretensioner may pull the belt back to move the occupant's upper torso back into an upright position. The harmful levels apply, until the belt ruptures at a point of maximum stress (although some belts are not designed to fail, so there will be continued restraint during the entire crash sequence). The effect of the pretensioner is to move the occupant's head backward enough to provide additional room for its subsequent forward motion. With a pretensioner, the forward horizontal head movement may be cut in half, thus changing where the head might impact, its force vector, and its relationship to a quick inflation and deflation of an airbag. There may be deployment of a knee bolster to alter the lower torso movement to prevent submarining under the instrument panel, and to help keep the occupant's torso upright during the crash sequence. The energy-absorbing steering column may collapse if the tilting steering wheel and direction of impact of the upper torso are in alignment. Of course, there may be airbag deployments that cushion and return the head, before contact with hard objects such as the steering wheel and B-pillar. The collision may be from the side, but many protective devices were designed primarily for frontal impacts at about 30 mph full barrier. The collision may be from the rear, so the seatback may have a significant tilting or horizontal movement to the rear and the headrest (passive or active head restraint) may or may not be in an effective position relative to the occupants' pre-impact position. There may be flying objects such as loose tires, tools, boxes, or canned goods (groceries). See Figure 11.1 for an illustrative crash sequence.

The crash-pulse information may suggest why certain damage appears in the vehicle and the force vectors being applied to occupants (biokinetics) that induce or produce specific injuries. There are specific issues usually derived from the crash event sequence

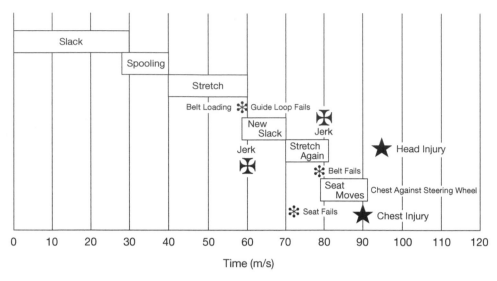

Figure 11.1 Crash sequence

diagram. For example, when did the seatbelt webbing fail or rupture and under what forces? Did the belts provide the expected benefit (occupant restraint) before they failed? Did they prematurely fail or tear apart because of a deteriorated or original understrength condition? An essential caveat is that each collision and vehicle may result in a different detailed time–event sequence, depending on where and how the vehicle was impacted and whether the occupant was upright, slouched, turned, or out of position (OOP).

The collision event sequence might be compared with what occurred in early design from sled testing or from full vehicle crash testing of exemplar or similar vehicles. There may be significant differences in preproduction tests using an average-size dummy or subsequent testing by various public or private entities. It may be necessary to determine the effects of vehicle repairs, modifications, accessories, environmental factors, and unique vehicle movements such as rotation, rollover, and secondary impacts. Gradually, a clear understanding of what happened and what caused the injury or damage will be revealed.

(3) Black-box data

Many vehicles have collision recording devices, a crash data retrieval system, and a data link connector. Data from such a system is often referred to as black-box data. It is similar in concept to aircraft flight recorder systems. It can provide pre-collision and crash information in an encoded form that can be translated for use (Rosenbluth, 2000).

The crash data retrieval system (CDR) may store data in the airbag sensing and diagnostic modules (SDM). Such data may include vehicle speed, engine speed, percent throttle, and brake switch circuit status in 1 second intervals for 5 seconds before the crash. Some modules include seatbelt usage, malfunction dashboard indicators (lights), and passenger airbag disablement.

Other black-box systems use the electroncially saved data for diagnosis and repair. The data may pertain to airbags (SRS), antilock braking systems (ABS), automatic traction control (ATC), cruise control (CC), engine fuel management (EFI), and seatbelt tensioners (ETR). The information is available in a read-only memory, in diagnostic trouble codes (DTCs), and is downloaded from the vehicle electronic control unit (ECU) by use of a repair level or engineering level scanner or microprocessor interface for one or more system units. For example, vehicle velocity conversions into miles/hr or km/hr could reveal speed changes before and during the crash sequence.

The accuracy of the data may be open to question. For example, the longitudinal accelerometer used to generate post-crash delta-V graphs may have to be corrected for side-impact vectors if known. Thus, the availability of such data does not mean the collision scene patterns of debris, tire or skid marks, crush deformation, witness statements, and other information are not necessary or only corroborative. Each information source has its own value and standing in terms of credibility for a particular situation.

(4) Momentum and energy

In a collision, momentum (*mv*) remains the same pre-impact and post-impact. Thus, momentum equations can be used without considering the damage to colliding vehicles or considering energy relationships. The kinetic energy (KE) or energy of motion does change as a result of a collision, because much of the energy may be consumed in creating the damage (crush). These facts are used for the energy or damage method, as contrasted with the momentum method of analysis.

There is a variety of techniques to determine pre-impact speeds, impact speeds, changes that occur during collisions, the exit speeds and directions after vehicle disengagement, and the final resting points. The time-based history of collision events may result in an illustrative graphic printout depending on the software used in a computer-generated analysis. In general, these calculations are based on momentum and kinetic energy equations (see pp. 153–4). The factual information necessary for calculations could include the mass of the vehicle, its wheelbase, degree of braking, the damage location and contour, the direction of deformation, rotational inertia, and intervehicle sliding. It may include a spin analysis and vaulting analysis. Different specialists may have differing opinions on whether energy calculations are better than momentum calculations, so both are generally advisable.

The computer models must often utilized by accident reconstructionists include the following.

- **CRASH** (Cornell Reconstruction of Automobile Speeds on the Highway). There is also CRASHEX (CRASH Extended), WinCRASH, and m-crash.
- **SMAC** (Simulation Model of Automobile Collision). There is also M-SMAC, the McHenry Version of SMAC, and WinSMAC from AR Software, Redmond, WA.
- **IMPACT** (Improved Mathematical Prediction of Automobile Collision and Trajectory).
- **TBS** (Simplified, Interactive Simulation for Predicting the Braking and Steering Response of Commercial Vehicles).

There are other models such as HVOSM and AITools-Linear Momentum. Based on the SMAC model, there are trademarked programs such as EDCRASH, EDCAD, and EDVAP, by Engineering Dynamics Corporation, Beaverton, OR. Other companies and consultants have modified models or personal models.

There are *rollover animated models* that show vehicle motion, occupant motion, restraint system function, and impacts. There are *acceleration recorders* to measure acceleration events and force–time acceleration wave forms, during either vehicle handling or crash testing. In fact, there are many other tools of the trade that can be used by accident reconstructionists to solve almost any issue.

(5) *Injury classifications*

Understanding the injury may become a vital part of an accident reconstruction, since specific injuries may lead to inferences about injury causation and product deficiencies. For example, a leg fracture may suggest that the leg was flailing about during the collision, the lower torso submarining under the instrument panel, or the inward crushing of the vehicle that compromised the leg space in the interior of the vehicle. Similarly, a lower spine (lumbar) injury may suggest a violent pitch-down of the vehicle during rebound and a possible problem with a metal or plastic pelvic restraint in the forward position of the seat cushion area.

In general, the location and magnitude of the *specific* injuries are the prime focus, not some general understanding or method of injury classification. However, commonly accepted or consensus methods of evaluating general injuries do have some value for company activities related to injury pattern and trend identification (the cumulative magnitude of injuries), pareto (worst injuries first) studies, and for cost–benefit evaluations comparing actual injury costs and possible design improvement cost–benefits.

General injuries may be rated or assessed as to their severity by various techniques as suggested by the following acronyms that may be found in accident records.

- **GCS** (Glasgow Coma Scale, used for head injuries based on eye opening, verbal response, and motor response).
- **CRAMS** (Circulation, Respiration, Abdomen, Motor, and Speech).
- **IPR** (Injury Priority Rating, an impairment assessment based on mobility, cognitive, cosmetic, and pain categories, also levels).
- **AIS** (Abbreviated Injury Scale, used to classify injuries into one of seven categories). If there is more than one injury, an Overall AIS (OAIS) classification procedure may be used. A zero indicates no injury and a 7 indicates it is virtually unsurvivable.
- **ISS** (Injury Severity Scale, assesses the synergistic effects of multiple body injuries).
- **HIC** (Head Injury Criterion, discussed in detail elsewhere in this book).
- **HARM** (a method of combining injury levels and assigning economic values to the injuries).
- **MISS** (Modified Injury Severity Score).
- **RTS** (Revised Trauma Score). A trauma score used for rapid assessment and treatment prioritization.
- **TRISS** (Trauma Score plus ISS, used to assess injury to lower-age individuals).

There are other methods in use, but they do not utilize the results of detailed neuropsychological testing, the findings of neurosurgery, or the subsequent neurological treatment, recovery, or residual condition.

(6) Basic terminology

It is important to have a basic understanding of the relevant terminology when reviewing investigation, reconstruction, or police reports about an accident. Some *key definitions* (in commonly used non-technical language) are as follows.

Delta V

When two vehicles collide, there may be a change in the direction of travel (pathway) of each vehicle and a change in speed. Together, the speed and direction is called *velocity*. In a collision, the change in velocity of a vehicle is called *delta-V* (ΔV). There is a relationship between each vehicle's delta-V, its initial and post-impact momentum, and the energy lost or transferred in the collision.

Impulse

Each vehicle may be subject to a force from another colliding vehicle. That force may be exerted for a certain time duration. The magnitude and duration of the force, specifically the area under the force–time curve, is known as an *impulse*.

Contact time

The duration of force applied in a collision is generally 0.1 to 0.2 seconds and is known as the *contact time*. The maximum force is generally about two or three times the average force exerted during the contact time.

Momentum (p)

This is the product of the mass (m) times the velocity (v) of the center of mass of the vehicle. The *change in momentum*, of the center of gravity, is equal to the *impulse* (m) applied to it.

$$p = mv$$

Mass (m)

This is the weight divided by the acceleration of gravity or $m\text{-}w/g$. That is, weight divided by 32.2.

Rebound

The surface of a vehicle will rebound or elastically move back in the opposite direction of the forces first applied. The rebound is about the same as the relative velocity in the initial impact (less considerable energy lost in metal crushing and sliding friction). Note: differentiate rebound from restitution and the rebound in suspension systems.

Force (F)

A vector (magnitude and direction) capable of accelerating a mass:

$$F = ma$$

Acceleration (a)

The time rate of change in velocity. A *jerk* is the time rate of change in acceleration. Expressed as feet per second per second.

Kinetic energy (KE)

The ability to do work due to the motion of a mass:

$$KE = \tfrac{1}{2}\, mv^2$$

(7) General terminology

Of some value are the *definitions* of words and abbreviations commonly used in accident reconstruction documents, expressed in a simple non-technical manner. The following are some of the most common.

Antilock brake systems (ABS)

The use of a microprocessor, electronic sensors, and hydraulic components to automatically pump (apply) the brakes very rapidly and repeatedly to avoid wheel lockup, tire skids, and to achieve maximum braking efficiency.

A-pillar

The roof supports at the windshield. Other roof supports are the B-pillar (between the front-door window and the rear door window of a sedan), the C-pillar at the rear of the windows, and the rearmost is the D-pillar on some vehicles.

Ball joint

The flexible ball-and-socket joint used in front suspensions.

Beltline

The line formed around the vehicle's body defined by the lower edge of the vehicle's glass window panels.

Bobtail tractor

A truck tractor without an attached semitrailer.

Brake modulation

The varying brake-pedal pressure needed to hold the brakes in full engagement, but less than that needed for brake lockup, which could result in reverted rubber skidmarks (lesser braking efficiency).

Camber

The tilt in the wheels inward (negative camber or top tilt-in) or outward (positive camber or top tilt-out).

Caster

The amount (angle) that the wheel pivot axis is located ahead of where the tire meets the ground, so that the wheel trails behind the pivot and tends to self-center. More specifically, it is the longitudinal inclination of the steer rotation or kingpin axis. A positive caster exists when the steer axis (ground intercept) is ahead of the tire contact (the center of the tire ground footprint).

Center of gravity

The point in a vehicle about which it is in perfect balance regardless of how it may be turned or rotated.

Chassis

The parts of a vehicle attached to a structural frame (rails) or, for vehicles with unitized construction, everything but the body of the vehicle.

Class 8

The highest category of heavy trucks based on gross vehicle weight. Class 7 is medium-duty.

COE

Cab-over-engine truck.

Compliance

The resiliency of suspension bushings that cushion road bump shock, with some rearward motion but no lateral movement during cornering. Affects harshness of the ride and vehicle handling.

Contact patch

The circular or elliptical area in which the tire tread is in contact with the roadway or ground.

Cruise control

A device in which an electronic controller monitors the vehicle speed as compared to driver's set speed and activates the throttle linkage to maintain the desired speed.

Derivatives

Vehicle models using the same platform.

Drive-by-wire

The elimination of mechanical components between the driver and the vehicle. The use of electrical and electronic signals reduces mechanical linkage routing problems, wearout potential, and vehicle servicing requirements.

Eye ellipse (eyellipse)

A side-view representation of the scatter of eye points of vehicle drivers.

Fatigue fracture

The progressive growth of cracks in metal, from bending cycles and intermittent varying stresses, until the metal fractures completely. Usually manifested by progression, beach, or clamshell marks.

FCIS

Front cab isolation system, a vibration-reducing suspension system for trucks.

Fifth wheel

A coupling device attached to a truck tractor chassis that can pull a semitrailer and support about half the semitrailer weight.

Floorpan

The metal stampings that form the floor and serve as the foundation for the vehicle's mechanical parts. It serves to fix the dimensions for many structural and external panels.

FMCSR

Federal Motor Carrier Safety Regulations (USA).

FMVSS

Federal Motor Vehicle Safety Standards, published by the National Highway Traffic Safety Administration (USA). May be subject to revisions and serves as a minimum standard for the areas specifically included.

Gasoline (petrol)

The petroleum hydrocarbon fuel used to power internal combustion engines. May contain extra butanes and pentanes in the winter, inhibitors of oxidation and gum formation, antiknock agents, surficants, and dyes.

Gravity acceleration

32.2 ft/sec².

HIC

Head injury criterion.

H-point

The point of rotation of the hip joint of an occupant seated in a vehicle. See SAE Standard J826 for specification details.

HVAC

Heating, ventilation, and air conditioning system (climate control).

IC

Integrated circuit (electronics).

ISE

Initial impact speed estimation.

LED

Light-emitting diode.

LP

Liquid propane gas (fuel).

Macro

Visible to the human eye.

Metallography

The process of polishing the face of a fractured metal surface, etching it, and examining the microstructure under optical magnification of about 1000×.

Micro

Visible by optical microscopes.

MMI

Multimedia interface (telematics).

OEM

Original equipment manufacturer.

Oversteer

A vehicle handling condition in which the rear of the vehicle swings wide or loose.

Pitch

Usually, the upward or downward movement of the rear of a vehicle about its longitudinal axis. The angular velocity is about the side (left to right) axis of the vehicle.

PLP or PRP

Principal locating point or principal reference points. Used to measure the deviation (location) of vehicle members from a standard reference point to determine assembly dimensional errors or the extent of collision deformation.

Prime mover

The tractor that tows or pulls trailers, construction equipment, or farm implements. That is, the self-powered vehicle or power unit for propulsion.

RDS-TMC

Radio data service – traffic message channel (Europe).

Ride-down

When a vehicle occupant is restrained and moves in conjunction with the vehicle, the occupant rides down the crash deceleration rather than being in free flight. The occupant experiences no spikes or jerks separate from the vehicle under ideal conditions.

Ride steer (bump steer)

A condition in which a wheel steers slightly with changes in suspension extension and compression.

Roadholding (lateral acceleration, cornering limit)

The vehicle's grip on the road. Measured in *g* levels achieved at the maximum speed at which a vehicle can negotiate a given curve.

Roll (sway or lean)

The rotation of a vehicle body about its longitudinal axis. Occurs when the vehicle's center of gravity is higher than the rotation axis. It may be sideways turnover of the vehicle. The angular velocity is around the longitudinal axis of the vehicle.

Roll steer

Steer angle changes due to suspension roll.

ROPS

Rollover protective structures that are used to protect occupants. May be a safety cage for construction vehicles such as bulldozers. Used for protection during side-over-side or end-over-end tipovers and rollovers.

SEM

Scanning electron microscopy.

Slack adjuster

The lever that serves to increase the pushrod force on the brakes of a truck.

Slap

Noise from tires traversing expansion joints, tar strips, and road seams.

Speed (v)

The distance traveled per second or hour, e.g. 60 mph (88 feet per second or 96.54 km/h).

Sprung weight

The vehicle weight supported by its suspension system.

Square wave crush

A vehicle deceleration (time–force) curve of fairly constant magnitude.

Steer angle

The difference between where the vehicle is headed (velocity) and the wheel plane (wheel direction).

Stress

Force per unit area.

Target vehicle

The vehicle that is impacted or targeted.

TEM

Transmission electron microscopy.

Thump

Audible periodic sound generated by a tire.

Toe-in and toe-out

If the forward portion of the wheel is turned inward it is a toe-in; if outward it is a toe-out.

Torque

The product of a perpendicular force acting over a given distance.

TPO

Thermoplastic olefin composite plastic (interior trim). Used for high-impact performance for head impact, air bag deployment, and B-pillar covers.

Traction

Traction is a measure of the force derived between a tire or tractive device and the roadway or other medium in which the tires operate. The difference between gross tractive force (theoretical) and net tractive force (actual pull) is the towed force.

Trajectory

Path of motion.

Transfer case (second transmission)

The transmission component that permits a vehicle to go to or from two- to four-wheel drive.

TSP

Telematics service provider.

ULF

Ultra low frequency radio waves.

Understeer

A vehicle-handling condition in which the vehicle resists turning and wants to continue in a straight line. It is a slip angle problem of the front tires as compared with the rear tires.

VHF

Very high frequency radio waves.

Wheel hop (axle tramp)

Lifting of a tire from full contact with the road, usually the right rear wheel under power.

Wheel stops

Concrete or metal chocks used as stops for vehicles in parking areas. May damage vehicle undercarriages, permit excessive bumper overhang, result in one-wheel tire contact, and serve to trip pedestrians. Contrast with continuous curbing and guardrails.

Yaw

Usually, the sideways movement of the front of a vehicle. It is the angular velocity about the vertical axis of the vehicle.

Yield stress

The stress level at which plastic deformation is initiated.

(e) Reports and graphics

An important decision is how to report the accident reconstruction and its derivative findings. It may sound simpler and easier to make an oral report or have an off-the-record discussion with a responsible design group. Sometimes only a brief conclusion conveyed to a supervisor is necessary. The objective should be to preserve a product's history because design engineers who learn about the problem may soon retire, leave work, or be transferred to another job. All too frequently, new engineers repeat the same old problems. A written report, oriented toward product improvement and intended for future model production, could serve a valuable experience-retention function.

The very process of preparing a written report generally requires greater attention to detail, accuracy, and conformity to recommended practices. It should be prepared with the company record-keeping policy and who the potential readers of the document might be in mind. For example, how would it look or be interpreted if the report eventually emerged before a court, a legislative committee, a regulatory agency, or a competitor seeking market advantage? In this regard, there should be completeness, adequate factual foundation, an avoidance of speculation, and cautious use of words. Parsimony may be preferable.

Is there a way in which there could be a very quick transfer of information, assuming the reconstruction has identified a specific defect and suggested a silent recall or some other means to correct the problem? Could the communication be done without possible adverse publicity that could cause harm to a brand image?

The accident reconstructionist may have the computer software to generate and print accident scene diagrams that are simple or complex, black-and-white or color, and static or animated. There are many companies that specialize in realistic 3-D animations that are quite effective, but sometimes expensive. There may be photographic enlargements, medical illustrations, timelines, charts, graphs, document cleaning and highlighting, and the use of human digital modeling. The ultimate question is how best to communicate to a target audience the significant facts and conclusions in a fast and convincing manner that promotes action where warranted. This also means avoidance of an uninteresting and complex reading burden on others that might result in avoidance, dismissal, or forgetfulness of the results.

(f) The follow-on

It is important that the manufacturer be informed of safety problems discovered from accident investigation and reconstruction. It is not enough to inform the company's insurance, legal, reliability, quality, warranty, marketing, or their testifying experts' department. The reason is simple: the responsible engineers must know if the safety problem is to be corrected on existing products and prevented on the new products being designed or that will be designed. The responsible engineers may coordinate with those identified in the company's policy manual or organizational responsibility charts as being ultimately responsible for overall product safety. This coordination is to secure a range of options as to what to do and how to do it, to assure that there is management approval, and to act in accordance with financial and public relations constraints.

A periodic follow-on effort is needed for a company to benefit from the in-field accident reconstruction activities performed by any person or group. This is because of the following five factors.

(1) Isolation

It is not unusual for the design engineers to be isolated within the company, and to be focused almost completely on new product design. The old products may be relegated to the less experienced. They may believe that their personal responsibility ends with the approval of their final design. For example, several military helicopters crashed, and accident investigators traced the problem to where some electrical lines that passed through a hole in the bulkhead became frayed and short-circuited. The problem received considerable publicity. About a year later, during a visit to the company, the problem was mentioned to the two design engineers responsible and their supervisor. The expressions on their faces and verbal remarks indicated complete surprise. Nobody had told them!

(2) Forget and repeat

It is not unusual in the automotive industry to discover a safety problem, resolve it for the next few models, then have it reappear five or 10 years later. This may result from the transfer, replacement, advancement, or retirement of those who hold the relevant knowledge. New engineers may not benefit from the historical experiences stored in the memory of former employees. In essence, there may not be a formal system of experience retention.

(3) Not proven

It is very easy to assume a personal and company defensive posture, to tell others 'prove it to me', and thus not assume the additional work involved in accepting a negative finding of a possible safety defect. The proof required may vary from 'just a little more' to a burden that approaches the 'impossible'. A high burden may be perceived as protecting the company, manifesting an appearance of loyalty to the company and its engineers, and saving time and effort on a past generation of products no longer earning money for the company. It is an avoidance of personal blame for recognizing something negative, rather than perceiving the positive aspect or the need for action.

(4) Reputation

A public perception may grow that there must be a safety defect, deficiency, or problem with a product. But the technical or company perception may be that there is no such problem. The issue is not whether there is a real problem, because public opinion and customer belief is, in fact, the problem. Unfortunately, public opinion generally trumps technical opinions. In such situations, something substantial must be done to protect the reputation of the product, brand, or company. For example, a distributor or seller may provide poor service, overcharge, misrepresent, be less than honest with the customer, and have negative interpersonal relationships. In one such situation, the manager indicated that this was just typical marketing promotion, with a focus on sales, and that some exaggeration or puffing always occurs. He remained silent about poor service operations and deviations from prescribed repair procedures, as if there was nothing wrong with what was occurring. Factory representatives felt otherwise. For them, product reputation and market share was at stake.

Some companies have spent a great deal of money attempting to deny a safety problem, because it was embarrassing to their design engineers and in conflict with some representations made to the customer base and government agencies. If denial is done, an equally important effort should be made to remedy the causes of the adverse public relations by what appear to be positive and constructive actions. Adverse opinions often grow into long-term difficulties, so follow-on activities need to be equally long-term in character.

Engineers involved in accident investigation and reconstruction sometimes have a basic ethical conflict between client confidentiality and their moral, sometimes legal, obligation to disclose dangers to the general public. The engineer may discover a safety problem, but what should he do? One rational approach to disclosure is to engage in reasonable dialog and follow-on efforts to encourage remedies to substantive safety problems. The proponent of change should be aware that you can't fight city hall unless you are very diplomatic, convincing, persistent, and willing to listen to the other side.

(5) Previously overruled

Many companies perform costly advanced engineering studies lasting months or years for proof of concept, technical feasibility, pricing, compatibility, and possible benefit to future products. A particular study may meet all criteria, be approved, internally coordinated, and actually published as a company design study. Then someone in the chain of command or coordination procedure may state that it is not presently needed for marketing, that it is disapproved pending further study, or that it should be 'stockpiled'. Once overruled or placed on the shelf, such studies are rarely resurrected, since others may believe that there could be some unknown reason for the non-acceptance. An accident investigation and reconstruction with a recommended remedy may identify a need, but an earlier overruled design study may have been filed, long forgotten, and provide a remedy already evaluated. Where appropriate, the company accident reconstructionist or recipient of his report might review prior design studies to determine their applicability and possible use. It may save time and money by avoiding an essentially duplicate study. Thus, the follow-on effort may include activities extending beyond a simple reminder that something needs to be done.

12 Special design problems

(a) Age restrictions

(1) Motorcycles

A half-sized electric motor-driven motorcycle was sold in a large department store located in a heavily populated urban area. It was marked 'not for use' by persons over 90 pounds in weight. Converting this weight limit of 90 pounds (40.82 kg) into an age limit, it would be 12 years (50-percentile), 15 years (5-percentile), or 9 years (95-percentile). That is, the age limits would be 9, 12, or 15 years old for that weight limit (with a conversion according to Snyder *et al.*, 1997).

Since these youthful ages are well below the motor vehicle driver's license requirements for that state, the implication is that the motorcycle would be ridden only on private roads. But how many children would remain off the public roadways if they were driving an attractive motorcycle with bright chromium-plated parts? How many would avoid driving the motorcycle into open fields and cross-country pathways intended for more conventional vehicles? That is, it is predictable that children, without licenses or training, would operate the motorcycle over rough terrain and among other more substantial vehicles on the open road.

The motorcycles were equipped with a 12-volt battery to power the electric motor. They were also equipped with training wheels. That is, inexperienced drivers were anticipated. In one display, no driving restrictions were apparent as an exemplar vehicle sat on the floor attracting the attention of youngsters accompanying their parents.

Is it the manufacturer's responsibility to clearly define the driver's age limits, both the lower and upper bounds? During design, what criteria were considered in terms of the driver's ability and skill level required to control the motorcycle, to use it properly in designated riding areas, and to avoid injury? If it is considered to be strictly a parental responsibility to determine the lower and upper bounds of driver age, should other factors be considered, including personality traits such as aggressiveness, sensory–motor coordination, decision-making prudence, intelligence, and the condition of available private roadways? How should this be communicated to the parents?

Without specifying clear age restrictions and the specific conditions of use, the ambiguities serve to induce safety problems. The question also arises of what unique safety devices are needed for the specific age group expected to operate a particular motorcycle.

(2) Replicas

One vehicle was 'recommended for ages 18 months to 4 years' for children weighing up to 50 pounds. It was a battery-powered, small-size, replica of a well-known passenger vehicle. The maximum speed was 2.5 mph, both forward and in reverse, with steering and a brake for 'quick stops'. Add-on stickers stated 'let kids customize their vehicle'. While it may be assumed that parental supervision may restrict the use of the vehicle to safe locations of a paved roadway or sidewalk, it is predictable that some young children would want to drive their customized vehicle to a neighbor's home or to explore on a roadway being traveled by other vehicles. The age restrictions given for the vehicle were apparently based on the vehicle's dimensions and load capacity, not on the child's judgment or actual knowledge of the risks of driving the vehicle on various paved surfaces. The vehicle was advertised as a Christmas gift for children. Other replicas of different full-size vehicles are also on the market.

(3) Other vehicles

Similar problems as to age restrictions or limitations may apply to ride-on lawnmowers in residential areas, various types of all-terrain vehicles (ATVs) used for recreation and work tasks, snowmobiles used in forested areas, and even on a one- or two-person water jet ski used on a river or canal. There may be more of a potential safety problem on remote farmlands where children of a very young age must perform work chores and learn how to use farm equipment.

Foot-powered scooters became popular in 1999 and 2000 and this led to scooters powered by small gasoline engines or electric motors. Children 12 to 14 years of age were observed riding their powered scooters in heavy urban traffic. In 2001, one electric-motor-powered scooter, rated at 9 miles per hour speed, costing less than $200, advertised an 'optional capacity up to 150 pounds'. There were no age restrictions.

At the same time, a battery-powered 'electric scooter' was advertised as having a 12 mph maximum speed and intended 'for ages 16 and older'. There was a caveat: 'Note: Check with local authorities for street use regulations'. That is, street use was expected and unlicensed drivers were warned. But consider what might happen to a 48-pound vehicle in a traffic mix of passenger cars, SUVs, light trucks, and tractor-trailers.

A three-wheel, one-occupant, full-sized plastic bubble-type vehicle was advertised as having a well known gasoline-powered motorcycle engine. It was also indicated that the vehicle could be insured in a motorcycle classification. It was intended for road use, but clearly lacked basic crashworthiness features.

The adult-sized two-wheel gasoline powered 'motor scooter', capable of seating two riders on a banana seat, had a speedometer marked up to 80 mph. The scooters were displayed at the retailer and were commonly used *without* a windshield. The windshield contained a warning indicating:

- speeds above 40 mph may create stability problems
- cross-winds may cause stability problems.

The dealer-supplied helmet did not contain a visor to protect the eyes. It is not to difficult to think about a highway use, where trucks or buses create cross-winds and dust, so the driver may be temporarily blinded while the vehicle is experiencing stability problems. It may be argued that the driver should use appropriate safety equipment, has been made aware of the vehicle's stability problems, and drives at his own risk.

(4) Boards

An 'electric powered board', capable of a 16 mph rated speed, was sold in discount department stores. As sold to the customer, it was contained in a large cardboard box containing the following warnings:

- should not be used by minors without adult supervision
- not intended for operation on public streets, roads, or highways
- do not operate this product in traffic or on wet, frozen, oily, or unpaved surfaces
- do not operate under the influence of drugs or alcohol
- should not be used by persons without excellent vision, balance, coordination, reflex, muscle and bone strength, and good decision-making capabilities
- does not conform to motor vehicle safety standards
- user assumes all risks.

Each of the warnings seems to be a recitation of what may be expected in terms of safety problems. It would be difficult to use the vehicle and avoid the exclusions stated in the warnings. The warnings appear to be from a lawyer-analysis rather than an engineer-analysis.

For the design of vehicles that are capable of use by youngsters, the lessons learned from the introduction of other vehicles should be carefully studied. This includes 3- and 4-wheel all-terrain vehicles, snowmobiles, small racing scooters, and the ongoing experiences of the vehicles described above.

For additional information on all-terrain vehicles, see Karnes *et al.* (1990), p. 7.

For age-related problems on agricultural vehicles, see Saran (1988), p. 653. For age-related problems on mopeds, minibikes, and motorized bicycles, see Brown and Obenski (1988), p. 71. For additional information on warnings, see Chapter 5.

Unpowered skateboards and scooters may achieve a maximum speed of 11 to 13 mph (18 to 21 km/h) on smooth level asphalt roadways for children of 11 to 13 years of age (Felicella, 2001).

(5) NEVs

Neighborhood electric vehicles (NEVs) are golfcart-type vehicles that may lack horns, doors, bumpers, and airbags. In the United States, their use has been approved on public roadways that have speed limits up to 35 mph. Yet many have not been crash-tested at even 30 mph because of an obvious lack of structural integrity. They are competing for road space with large SUVs and trucks. Some do not travel at greater than 25 mph, so they may be the slow vehicles that induce accidents. Some of these small low-speed electric vehicles have no side protection, so the occupant could come into direct contact with any impacting vehicle. Their market availability has increased as a result of credits earned for zero-emissions requirements for total vehicle sales within a state. There is a counterargument that they should be compared with motorcycles rather than conventional passenger vehicles. However, motorcycles have an accident fatality rate 25 times that of passenger vehicles. The NEV use could be confined to golf courses, for designated off-road or private road situations, or in appropriate community or neighborhood use with restrictions that limit the speed and size of other competing vehicles on the roadway. They are also the simple type of vehicle that would be attractive to young children as passengers or those learning to drive. They may be more accident-prone, because of the drivability problems, for both the underage and the elderly driver. However, with design improvements, future NEVs could provide a very small vehicle classification in the current range from premium and luxury vehicles to small budget and economy models. A still lower classification might include the two-wheeled electronic scooter which is directed by the motion of the rider. Such new classifications require special attention to age restrictions for the driver.

(b) Entrapment

A mother decided to leave her infant in the automobile while she made a quick run into a department store. The vehicle interior had been cooled by air conditioning, she left a crack at the top of a window for fresh air, and then she locked the vehicle doors. The outside (ambient) temperature was about 95°F and it was a nearly cloud-less day. After she left, the temperature quickly rose to the point at which hyper-thermia or heat stroke could occur. In 20 minutes the vehicle interior temperature was 122°F (50°C), and in 40 minutes it was 150°F (65.6°C). For purposes of comparison, the human body's normal oral temperature is 98.6°F (37°C) and a rectal temperature beyond 104°F (40°C) may produce hyperthermia, heat stroke and death. Hyperthermia indicates that there is insufficient body heat dissipation and it can result in discomfort, an increase in oxygen consumption, reduction in mental acuity, production of delirium and stupor, and finally death may occur.

Fortunately, some passersby alerted a guard about an infant in the hot car and when the mother returned to the vehicle she was arrested for child endangerment and negligence. This scenario occurs regularly due to unawareness of the hazard or forgetfulness. It occurs to children, pets, and impaired older adults. It is a human error that sometimes results in a loss of life.

Another form of entrapment occurs when a child climbs into the interior cabin or trunk of a vehicle and becomes locked in. The child may climb through the pass-through space between the vehicle interior and the trunk. A victim of criminal behavior may be bound, gagged, and left in the trunk. The closed trunk is dark, small, has no fresh air, and may reach high temperatures.

Some vehicles have internal release mechanisms such as a pull-handle or touchpad inside the trunk that might trigger a distinctive sound on the automobile horn or unlock and open the trunk. However, this requires logical thought processes, even if the release mechanism is marked with florescent paint. Design efforts are underway to perfect a presence-sensing detector (for movement and heat) for the trunk, a carbon dioxide sensor to monitor breathing, a heartbeat sensor (vibration), an infrared or radar automatic trunk release system, and a sun-powered or low battery drain air blower (timed cutoff) to reduce the interior heat in a parked vehicle.

The entrapment problem may be exacerbated if children are provided with ride-in toys that resemble various types of vehicles, because they are tempted to move over and into the real automotive vehicles and experiment with pushbuttons (such as window closure or door locks). Interior temperature control may be needed in desert areas with high daytime temperatures for the protection of humans, groceries and other cargo, and for the interior of the vehicle itself.

There is also another type of entrapment: that which results from a collision. The doors may jam, the metal may crush, and the occupants may need to be cut out with the jaws of life. This assumes that the rescue team has been alerted by a wireless cell telephone, the Internet, or an emergency telephone number and knows the location by map or global positioning system. Hopefully, there will be emergency medical services available and quick transport to appropriate health care or trauma centers.

Rescue is a design problem. Buses may have escape panels on the side and rear, and escape hatches in the roof. They may be equipped with small pick-hammers to break the tempered glass windows. In passenger vehicles, windshields made of laminated glass do not permit easy egress, but side windows of tempered glass may be shattered into many harmless fragments and permit clean exit. If the vehicle is fireworthy, there is additional time to escape or be rescued from collision entrapment. In essence, the need is for timely communication that an unfortunate event occurred, the pressing need for help, and the vehicle location. This should be followed by rescue (extraction) and transportation to medical services. But much depends on vehicle design in terms of crashworthiness, occupant protection, fireworthiness, rescuability, and escapability from entrapment.

(c) Ladders, steps, and platforms

Automotive vehicles increasingly abound with various forms of external steps, ladders, and platforms. Whether part of an original design or just an aftermarket access aid, many present unnecessary hazards and some violate 'good practice' design criteria well known in other disciplines. A few examples might illustrate some of the problems.

A double-deck automobile transport trailer had two ladders (one on each side of the trailer) to provide access to the second level. The ladders were located near the rear of the trailer and appeared to be original equipment. The distance between the rungs was 12 inches, which is customary for ladders in general use. The rungs were round, so that there was only a point (a narrow line) contact between the shoe of the climber and the smooth metal surface of the rung. Any significant lateral force exerted via the shoe sole would have little resistance and a slip could occur. The rung was exposed to rain and condensate, snow and ice, plus road dust and dirt. The shoes of the driver were exposed to the same slip-inducing elements. A short shoe slip might not seem serious, but any unexpected imbalance could topple a climber. Slips and falls are a high-frequency source of injury.

A concrete mixer had a ladder on the side of the vehicle, starting or ending more than 3 feet off the ground. The first step of the access was difficult, the last step coming down was a dandy! Similarly, a van had a makeshift aftermarket ladder at the rear, partially distorted because it was obviously understrength, with the first rung about 3.5 feet above the ground. One 'step bending test' consists of an 800 pound (362.9 kg) load applied to the center of the step by a device 3.5 inches (8.9 cm) wide. Casual observation suggested that the concrete mixer ladder could pass this test, but the van's ladder would fail the test.

There is another common test, involving a vertical sustained load of 300 pounds (136 kg) applied to a step or rung, but this does not account for the impact of a moving foot (dynamic load) of a person carrying some object (added weight). It does not meet the commonly used safety factor of 4 for the weight of a 95-percentile male. Certainly, external ladders are subject to damage and deterioration during their service life and, therefore, will lose some of their original strength. By these standards, a 300 pound test load would be inadequate. It illustrates the fact that some common sense should be applied when determining test procedures, so they will reflect real-life conditions rather than be an easy pass for substandard material. Hazard creation is easy; hazard prevention takes a little more time and effort.

Some early SUVs had floor levels high off the ground. Others were modified by suspension changes to be even higher. Access was difficult for some people. At first, there were no steps to help the driver and passengers climb into the vehicle. Subsequent models of various SUVs provided just a chromium-plated round tube as the first step, then a tube with what looked like a slip-resistant cover, then a short padded tube depression, and then a flat slip-resistant platform step. Some later models appeared with a running board to ease entry and exit from the vehicle. On some vehicles, the step width was so short that an average male's foot could not be inserted sufficiently to support the ball of the foot. Some were exposed to rain, icing, and accumulated snow, rather than be covered by a door extension or other built-in cover.

There are kneel-down buses for disabled people, but access may be difficult to some airport shuttle buses or vans. One nine-passenger shuttle had passenger access doors on one side with two long slip-resistant steps. Unfortunately, each step was about 2 inches wide and almost one on top of the other (3 inch risers). With practice, passengers might place their shoes sideways for support on the steps, but most passengers avoided the steps when entering or exiting the vehicle. The steps were too narrow and too close together for use by unfamiliar passengers. The treads could have been wider, with only one riser and step, and could have been powered if there was a need to retract them. The steps should have been supplemented with internal handholds to accomodate the varying dexterity of an unrestricted client population, some of them predictably fatigued and clumsy on arrival at an airport after a long trip.

On some vehicles, the first step is a depression in the rear bumper or a slip-resistant cover on the top of the bumper. A plastic trim on the rear bumper may be more decorative than functional, particularly if the rib lines (slip-resistant ridges) run laterally, not in both directions. Smooth lateral ridges may increase slip potential. Some large light trucks (pickups) require standing on a short running board, then placing one foot into a recessed step at the front of the bed of the truck. Whether a person is supposed to stay and stand on only one foot, or go higher is ambiguous at best.

Within some vehicles such as single- or double-decked buses, there may be interior steps, wells, or stairways to the upper levels. The vertical distance (risers) between steps may be reduced, tread nosings distinguished by colored paint or lights, and handholds or handrails provided. Provision should also be made for passengers carrying heavy luggage. Both the interior and exterior steps may have lights for illumination on dark nights. While slip-resistance may be attained by use of aluminum oxide particles in paint or mastics, deep ridges (a ribbed block pattern) may be more appropriate.

There have been many safety problems regarding access to truck cabs or 'cockpits', to the enclosed rear cabin of the truck, to the fifth-wheel control area behind the cabin, and to the interior of the truck cargo area from the front and side. A wide variety of OEM steps, ladders, and platforms can be seen on the highways. Many are aftermarket accessory devices and some are home-made climbing, working space, and descending aids. The trucker's rule is that at least three human contacts are maintained with the vehicle; that is two hands and a foot (while moving the other foot) or one hand and two feet (while moving the other hand).

The steps should be self-cleaning and draining, easily accessed, appropriately located, wear-resistant, with equally spaced risers, strong and braced, slip-resistant, and properly dimensioned.

A narrow and short toe-hold does not meet those criteria. A step that retains freezing water does not meet those criteria. A step that induces unsafe climbing behavior does not meet those criteria. A hanging stirrup step may or may not meet those criteria. A serrated expanded metal grip-step may be desirable for construction vehicles where sticky clay mud is a common experience.

Some light trucks have tubular racks for ladders, as factory optional or aftermarket accessory equipment. Ladders on such vehicles may be supported by rather thin round structures that could damage the fiberglass rails and also leave them exposed to ultraviolet radiation. Only a few have protective enclosures (slots, pockets, or supports) to reduce damage to the ladders.

An aerial work platform was extended on a long boom. An industrial safety engineer climbed up an extendable ladder on the boom. There were no side railings for sliding handholds or body restraint. When he attempted to descend from the platform to the ladder, he slipped and fell to his death. Transition areas may require special attention. There are long ladders on some fire trucks (aerial ladder trucks) for control of the water hoses or nozzles, to gain access to windows and roofs, and for rescue. The rescues are often dramatic and heroic events, but the firemen have special training, frequent rehearsals, and specific procedures to enable their safe use of the ladders.

There are many truck ladders that provide good ground and side clearance. They have bottom sections that can be pulled out, are spring-loaded to slide down, or can be folded down to reach about a foot off the ground. There are vehicles with powered steps that automatically lower when the vehicle stops. In essence, a well-designed ladder should have equal spacing between the rungs or steps and to the ground level. But the remedy should be commensurate with the risk. Style need not be sacrificed for functional safety; both can be combined.

There are many safety features and devices, invented during the past 100 years, used in various industries, designed to provide slip and fall protection (Templer, 1992). There are many trade standards with definitive criteria that could be utilized in any industry in the design of ladders, access steps, and work platforms. They should not be ignored.

(d) Batteries

In the past, automobile batteries were associated with many eye injuries from the splashing or expulsion of the acid electrolyte. There were claims about human error when the cell caps were removed for the addition of water (releasing flammable gas), the introduction of ignition sources (such as smoking cigarettes), and damaging the terminals by applying too much torque when reattaching the battery cables. Batteries have been a significant hazard: they can explode from the hydrogen gas that is generated during operation and recharging. This risk has been substantially reduced by the use of sealed batteries (maintenance-free batteries). Other design improvements include individual plate envelopes or separators to reduce vibration damage and collection of debris in the bottom of the battery, rounded corners on the negative plates to reduce electrical shorts between plates, heat-resistant plate compounds to enhance operation in higher temperature climates, and side-mounted terminals to reduce damage from overtorquing. These human error and design error risks took some time to be reduced.

As with many safety problems, design improvements were evolutionary in character, but they did occur.

As the electrical power requirements of vehicles increased, the battery output was increased from 6 volts to 12 volts, and now to a 48-volt system in some vehicles. This changed the character of the electrical shock hazard, since a higher voltage (force or power) meant that a spark can jump further, although it is the amperage (intensity or current) that causes the damage. But increased voltage generally means increased amperes. This hazard has led to increased electrical insulation, line or connector shielding, special warnings adjacent to hazard entry points, and owner's manual advisories. It also fosters a design emphasis on power management for the electrical systems, components, and the battery. But the myriad electrical and electronic devices now becoming available, as original equipment or post-sale accessories, suggest that special attention should be given to the derivative safety problems. For example, increased voltage may reduce lamp life and may increase electromagnetic interferences. In some vehicles, battery charging under varied load conditions and temperatures may be a difficult design problem. A 12-volt electrical system may range from 15 volts when cold, to 12 volts with maximum underhood temperatures, 24 volts from tow vehicle jump starts, 80 volt dump transients, and even to minus 13 volts with reverse battery conditions. Electronic components should have protective circuits and the capability of withstanding the varied voltages.

(e) Highway safety

One of the highway engineer's primary objectives or problems is to attempt to compensate for poor driving behavior. This is because driver error 'is typically stated to be a causative factor in 80 to 90 percent of all accidents' (AASTHO, 1973, p. 279). This includes the habits and attitudes of drivers, so the driver error problem is widely defined or of broad scope in terms of highway design.

Since the highways are designed to accommodate *all* drivers, this includes those impaired by alcohol, drugs, or emotional problems. The design features to compensate for driver error include adequate sight distances (stopping distances), pavement striping (lane marking), guardrails and median barriers, impact-absorbing devices and break-away supports for signs and lights, and features to prevent wrong-way access to the freeways. The safety features have evolved into a sophisticated technology. Of course, vehicle safety has to be considered both in terms of the better improved roadways and the less desirable roadways including the equivalent of off-road use.

Since pedestrians tend to be somewhat unpredictable, often taking the shortest distance between where they are and where they want to be, their walking speeds *should be considered* in roadway design. The speeds usually vary between 2.5 and 6.0 feet per second, with a design rate of 3.0 to 4.0 feet per second. This contrasts with the driver's reaction times. The driver *perception time* is the time required to appreciate a danger and decide to apply the brakes. It may range from 0.5 to 3.0 seconds or more. It is usually considered to be 1.5 seconds. The *reaction time* is the time required to apply the brakes. It may range from 0.5 to 1.0 second or more. It is usually considered to be about 0.75 seconds. There may be additional *human judgment* time in a complex roadway situation which may confuse the driver.

On highways in the United States, the posted maximum speed is about the 85th percentile of the speeds observed on the roadway. In other words, about 15% of the drivers are expected to violate the speed limit at any given moment. The violation might be considered driver error, but it is included in the design plan.

The interaction between the condition of the roadway and vehicle safety should not be underestimated. A large pothole can deliver a substantial shock to a vehicle suspension system affecting wheel alignment, it can damage tires by causing carcass breaks, and it can dent rims. The potholes may result from traffic that exceeds the roadway load capacity or from thermal cracking of asphalt pavement that permits rain water to seep down into the subgrade and soften or expand the foundation of the roadway surface. There are tests to determine predicted thermal cracking (such as AASHTO TP-96) and load-carrying capacity. The remedies might consist of overlays of increased depth and capacity, the use of appropriate specialized motor vehicles (milling, mixing, pavers, and rollers) for quality assurance, the use of pavement reinforcement mesh, and special care to achieve a pavement structure with the desired density. Roadways that deteriorate quickly and often are costly in terms of both roadway budgets and their effect on vehicles. Someone attempting to evade a large pothole, to avoid damage to his vehicle, may steer into another vehicle's path.

For further related information in this book, see Chapter 2, section (i), 'Positive guidance', and Chapter 7.

(f) The bulletproof office-on-wheels

At one time, heavy armored passenger vehicles were for presidents of countries who needed protection from the rare deranged assassin or from an attempted *coup d'état*. Then, lesser government officials and corporate executives residing in foreign countries needed protection from kidnapping and ransom demands, so the armored vehicles had to be capable of more evasive maneuvers, lighter in weight, and less obvious in appearance. They had to be bomb-resistant (to explosive mines) and bulletproof (to shotguns, AK-47s, and .44-caliber Magnums). They became light armored vehicles (LAVs) with communication devices such as cell phones and television. Some were stretched vehicles that could be used for client conferences in full security or secure offices on wheels. They also evolved into customized SUVs that are considered safe rolling offices that move from location to location. Because of the threat of terrorism, there have been requests for rolling fortresses that have tear-gas dispensers, door handles that can electrically shock, rotating license plates, devices that drop tacks and oil on the road, and even retaliatory weapons. The interior design of many include a full range of electronic communication devices and entertainment features. As with all customized vehicles, the production volume is comparatively small and funding is limited for special design studies. There is a balance between security and safety. Lighter weight glass and other materials are desirable. Greater structural strength, including the floor pan, is desirable. Special driver training is desirable. The use of convoys may be appropriate. Obviously, this is a developing technology to meet emergent needs.

(g) Pedestrians

In the past, projections from vehicles were not pedestrian-friendly. These included wheel spinners like Roman chariots in battle, large fins at the rear and large sharp fender edges on the front, a boxy overall shape, decorative emblems in the pathway of pedestrians thrown over the hood, and cutting edges on headlight assemblies. The vehicle design of contemporary vehicles is more rounded and pedestrian-friendly as a result of many studies of the human biomechanics after impact, over the vehicle, and return to ground level; being draped and carried by the vehicle; or being run over. But the human body is fragile, somewhat analagous to a set of plastic sacks of water, assembled over brittle bones, and held together by soft muscles and tendons. When this lightweight, fragile assembly is struck by the hard metal of a fast-moving heavyweight automotive vehicle, it is the human body that suffers the most.

In addition to a rounded vehicle shape, the vehicle's skin (sheet metal) should not have many hard spots, such as hood mounts, windshield wipers, and windshield washers. When a pedestrian is struck, the human body drapes or wraps around the impacting structure, then the body rotates around its center of mass, and the next impact is with the chest, shoulder, or head. This has resulted in annual pedestrian deaths of about 3,000 in Japan, 7,000 in Europe, and nearly 5,000 in the United States.

The design remedies under consideration include impact energy-absorbing bumpers and fenders, more space between the hood and the engine components, and airbags that might deploy above the bumper and over the hood, at the base of the windshield and over the top of the windshield, and elsewhere. The cowl and over-the-hood airbags require sensors that can predict impact and allow time for deployment. The current major design remedies are a smooth skin surface and sheet metal or plastic that is forgiving and energy-absorbing.

13 Future vehicle safety

(a) Introduction

There are many new features and devices being incorporated into the design of automotive vehicles. In addition, there is considerable promise of much more for future vehicles in the context of a rapidly evolving and competitive technology worldwide. Attractive marketing features may be found in terms of entertainment and information choices, occupant comfort options, and vehicle performance. However, those improvements that relate to safety are of special concern. Can the benefits be compromised by the end user? If so, in the future, will there be an effective risk reduction, hazard avoidance, or health promotion effort that could be undertaken by dealers, lessors, or vehicle owners?

This discussion centers on inappropriate use and the safety problems that can and have been created by what has been called improper use, misuse, abuse, misperception, lack of knowledge, and human error. It includes out-of-position or misaligned occupants during crash scenarios, improper seat adjustment, the improper use of optional safety devices or accessories, and common purchaser–user misunderstandings of risk-reduction practices. For future vehicles, there are the implications of universal design, sociotechnical analysis, and emerging precautionary principles to be considered.

(b) Misuse

The authors of this book recently conducted a study of nearly a thousand drivers in the United States and Canada. It revealed a widespread misuse of head restraints. The head restraint is an occupant protection device intended to keep the head and upper torso in alignment to prevent hyperextension and other spinal injuries. This study indicated that a third of all drivers, on major high-speed highways, had their head restraints at or below the shoulder level. Thus, their heads were unprotected from sudden rotation during rear and other collisions. Fewer than 20% of the head restraints were at an unequivocally safe height and this included many fixed-height seatbacks. This improper adjustment occurs despite instructions in the various owner's manuals to 'pull up' the adjustable head restraint as high as practical or at least to the level of the ears to avoid 'whiplash injury'. Misuse should be a major concern in the design of future vehicles.

(c) Out-of-position occupants

Even those with a high fixed-height seatback can defeat the safety objective if they are out of position. If there are bench seats, the occupants are free to shift their buttocks laterally so that their heads do not match the head restraint location. This may be to make room for others or to assume some more desirable seating position. But, even with deeply contoured bucket seats, some occupants will sit in a corner or lean away from the head restraint or the safe zone for the head. This may be to assume a more comfortable position to talk with others, to look to the rear out of the driver's window for approaching traffic, or to see and manipulate controls. Such behaviors may be sporadic or fairly constant, and with or without relevant knowledge as to why persons should avoid an out-of-position configuration. This should be explained to purchasers of new vehicles.

Some drivers may lean forward rather than keeping the head about an inch from the head restraint. Others may not keep about 10 inches distance from the location where the airbag may deploy. Similarly, some occupants still resist wearing seatbelts or intentionally create excessive slack. Some may attempt to sleep in the rear without the use of any restraint system.

Another out-of-position problem is the use of reclining bucket seats while driving. Despite the warnings in the various owner's manuals not to ride with the seat reclined, since the occupant could slide out from under the seat belt and be injured, many occupants use the recliner to rest, sleep, cat-nap, or gain extra room for a heavy body. Even when reminded, some occupants place their comfort over safety, and continue this practice or bad behavior. Should seats be capable of being reclined to a dangerous degree while driving?

(d) Human interaction

There are many areas of human–vehicle interaction likely to create problems if linguistic communication, multicultural habits, and subgroup propensities for error are not fully considered. Sophisticated areas such as collision warning and avoidance systems may be given considerable attention, but cargo storage (including groceries, tools and jacks, tires, gas cans, etc.) may be somewhat ignored. Intelligent vehicles still require skilled drivers. Hands off does not mean awareness turned off. Override has always been necessary. Future vehicles may have greater interaction capability, but what are the limits and controls to avoid greater risk?

Consider the interaction between an inflating airbag and a driver. The aiming point for the inflating airbag can be a tilting and telescoping steering wheel. The target sits on a seat, adjustable fore and aft, up and down, tilt, recline, with lumbar support, variable head restraint, knee bolster, and other parts of an integrated system. The human interacts and usually benefits despite a lack of technical knowledge.

(e) Distractions

There are increasing driver distractions such as the cellular or mobile telephone. In higher speed and crowded traffic, the driver acts as a single-channel information processor. Divided attention may mean distraction from or inadvertent loss of critical information in the visual field or in the focus of the mental concentration of the driver. A few localities ban drivers from talking on their mobile phones, but future vehicles will have advanced digital communication devices (including Internet readouts) with presently uncertain restrictions on use.

One of the authors was given a ride from an airport to an office about 50 miles away. The driver of the SUV was using one hand to drink coffee, then started to dial and use a cellular phone with the other hand, was steering with his elbows, and was weaving through high-speed traffic on an interstate highway. As a passenger, protests only seemed to encourage a macho ego performance.

We have seen television receivers, as an aftermarket device, located on the instrument panel. Even a computer screen, with a mouse, has been located for the driver's use and convenience. We assume that warnings were given about the distraction of using such devices while driving, but the temptation of such distractions is rather obvious. Voice activation and hands-free operation still involve similar safety issues. What, if anything, should be done to deal with the problems that might be created?

A business acquaintance in a distant city volunteered to bring one of the authors to a hotel. In the late afternoon the driver had three double alcoholic drinks, during the trip the driver urged that they stop for dinner at which time he had four double drinks, and upon arrival at the hotel the driver demanded another triple set of double drinks. He insisted that he was not intoxicated and could drive as he climbed into his vehicle for a quick trip home. The driver did not seem physically impaired, but surely he was mentally impaired? Such drivers constitute a significant subgroup in terms of accidents. Should they be driving? Did the alcohol serve as an attractive distraction for the driver during the trip or create a mental impairment? For future vehicles, what kind of devices would be reasonable to prevent the intoxicated from driving?

Whether something is an impairment or a distraction is debatable and superficial. What is needed is a greater depth of understanding to help devise safeguards to reduce accident propensity given new circumstances with future vehicles.

(f) Compensatory actions

The dealer has a unique opportunity for direct interaction with the prospective purchaser. An oversized or undersized driver can be shown seat adjustments for appropriate roadway vision and for head position in the airbag safety zone. Fortunately, motorized brake and accelerator pedal adjusters may help to locate the driver's head properly. If there is not a good fit, another vehicle model could be suggested. In other words, it is not just a sale, but a good fit. Enough time should be allocated for showing manufacturer's videotapes intended for risk reduction. The dealer can furnish a *personalized* instruction and motivation as future vehicles increase choices, options, and complexity for the consumer.

In a restaurant, we watched an aged disabled gentleman gradually work his way across the sidewalk, with the aid of a cane, thick eye glasses, and the help of his wife. Once he reached his automobile, he was able to lean on the car for stability until he could climb into the driver's seat. He gradually exited into heavy urban traffic. Incidentally, the restaurant had access for the disabled under the Americans with Disabilities Act, a 1990 civil rights law. There was design compensation for his disability at the restaurant, but what about his driving performance? For future vehicles, will there be any driving restrictions?

An elderly woman was attempting to renew her driver's license, but could not pass the eye examination. After loudly complaining that she only wanted a driver's license for identification purposes and that she needed a license in case of an emergency, the license examiner relented and she got a license despite her visual infirmity. What does this suggest for future vehicles?

Similarly, a non-English-speaking couple drove away and there was some concern about how they might decipher the road signs and the vehicle owner's manual. Should all road signs be pictorial and harmonized internationally, so problems would be reduced for foreign tourists?

We can assume that vehicle manufacturers will attempt to compensate or customize when it is reasonable, technically viable, and cost-effective. But a greater future burden may be cast upon less sophisticated dealers, purchasers, and users. The question is how to encourage appropriate actions by these subgroups to compensate for the inevitable technological improvements to automotive vehicles?

The 'informed purchaser' principle is that the dealer provides assisted learning and help in decision-making so that the driver understands what is being purchased and how to undertake appropriate risk-avoidance behavior.

(g) Universal design

The trend, particularly in world trade, is to achieve a better fit for a broader group of purchasers and users of automotive vehicles. The design-for-all concept of universal design requires greater knowledge of the representative groups in the proposed market-place. The knowledge is obtained by direct contact and this permits devising objective criteria to meet the customer's physical and behavioral needs relative to the product, i.e. proactive risk reduction by knowledge and action. Universal design should greatly influence the design of future vehicles.

Objective knowledge can be enhanced by the simple observation techniques that have been utilized to assess what is actually happening in terms of potentially harmful vehicle occupant behavior. Where possible, questioning or instructions have been used to gather further information concerning some of the reasons for bad behavior. This includes a simple failure to understand the consequence of the behavioral act, an understanding of the hazard created by the behavioral act followed by seeming disregard of the consequences, a momentary act of carelessness which involves a voluntary assumption of a perceived low risk, an intentional display of macho aggressiveness, or a defiance of the rules of good behavior or societal compliance. There may be problems of attitude, habit, custom, seemingly illogical thought, or a resistance to specific behaviors. A relevant behavioral (psychodynamic) analysis includes local customs, subgroup attitudes and beliefs, and customary practices. It may be supported by a sociotechnical evaluation. In essence, it is better to understand the marketplace in order to fully adjust to it in an age of rapid technological change.

(h) The precautionary principle

What is thought to be reasonable, safe, or necessary varies with time and place. Such relativity upsets notions of compliance with retrospective standards. Since a detailed and understandable objective standard of care is a necessary predicate to effective action, some appropriate criteria should be utilized for areas of scientific and technical uncertainty. The precautionary principle suggests that caution be exercised by the producer and risk minimized to the user where there is uncertainty: a simple warning flag where suspicion has been generated; risk assumption that is commensurate with the truth and certainty of the available knowledge. Where safety problems may emerge, the precautionary principle seems logical, although it is a heavily contested area of argument today. All trends suggest that this is an important criterion for future vehicles.

(i) Needed research

Assuming that future automotive vehicles will cause significant changes in the vehicle–driver interaction, appropriate research is now needed as a foundation for risk reduction efforts. The research that is needed to better understand future vehicle drivers should include all relevant attributes, variables and variances, and characteristic interactions. In addition to sociotechnical evaluations, it could include consideration of frontal lobe (brain) task management, working memory, rehearsal, executive functions, and other higher cognitive activities (Smith and Jonides, 1999). It should relate to the unexpected and uncertain, narrow the testing for meaningful and economic interpretation, utilize broadband representative populations, and assist in the ability to predict, foresee, and achieve meaningful marketplace opportunities.

It is not difficult to visualize a future automobile driver attempting to interact with a wireless mobile phone, computer, and Internet system that provides a steady stream of interesting e-mail messages, travel directions, stock quotes, emergency service, maintenance and service advisories, and a mix of personal and business-related search information. Is this to be accomplished by persons who can divide their attention with the driving tasks, is it to be done only when parked, or will there be some augmentation or sharing of all these mental tasks?

To what degree would this increase governmental responsibility for the training and licensing of drivers, improved roadway design, vehicle guidance, and traffic safety enforcement? What is the supporting research? To what degree could it be effective?

Dealers are important, since they may sell the financially important aftermarket accessories, perhaps do some conversions, actually see the vehicle purchaser, and are in a position to provide economical special instruction where needed. Where is the research on this topic?

In California, election (voter) ballots are available in seven languages and federal law requires multilingual voter information. Automobile vehicle owner's manuals may be available in several languages for the US market. For future vehicles, particularly in a global market, the language problem may be addressed in design by non-language pictorials with universal meaning and by the use of intuitive user-friendly tasks to be performed (Peters, 1991). However, the direct point of contact with the prospective vehicle driver is the dealer or lessor. The question is how can they quickly, economically, and effectively communicate unique or needed information about the vehicle in an appropriate language? Assume a foreign visitor at an airport auto-rental office leases an unfamiliar vehicle, with a new security system and some electronic 'gadgets' (such as 360 degree collision avoidance alarms, path follower system, and automatic braking and steering with override): what should be said and how, when a long line of anxious customers are waiting? What kind of research or application is needed?

In terms of universal design for future vehicles, further research may be necessary to enhance data gathered on the average male (SAE J885a; Stoffer, 2000). The intent would be to refine criteria relating to risk reduction in future vehicles. But this might result in the exclusion of certain subpopulations (such as oversized drivers) for certain vehicles. This, again, may be a dealer, lessor, and driver problem since they are in a position to take appropriate action and furnish the necessary guidance for the small proportion of exclusions.

In the future, vehicle assemblers may provide 'customized automobiles' on relatively short notice and with comparatively little delivery lag. This should not be confused with 'tailored vehicles' that are based on unique customer needs and provided by the dealers who are in personal contact with and aware of the special needs of the customer. For example, an overweight person might need a seat anchorage location moved rearward for driver comfort, pedal extensions for proper reach, and other adjustments for safe vehicle and accessory operation. To what degree should any standardized vehicle be altered, modified, or even sold to those with unique needs? What kind of procedures and criteria could be used in the dealer showroom to properly select or tailor a vehicle in terms of customer needs? This may be complicated by the independence of dealers, the habits and lack of knowledge of salespersons, and the effect on sales and profits. Some research indicates that too much should not be expected from the current dealer sales force in terms of either tailoring a vehicle or enforcing an exclusion of driver subpopulations.

Could there be a fundamental conflict between the general public's right or entitlement to drive a motor vehicle and a future vehicle's technical complexity and energy levels, the driver's cognitive choice loads, and the possible induction of user error? If so, what may be needed in terms of risk reduction for such vehicles? Again, it would be more economic and effective to base future risk reduction efforts on applicable research.

(j) The dealer's choice

Automotive engineers should have a specific objective in terms of the lifespan of a vehicle being designed and developed. This is needed to formulate appropriate test requirements pertaining to material aging and degradation, mechanical wear and durability, electrical reliability, resistance to adverse environmental conditions, and even abusive treatment of the vehicle. It may affect warranties and often has consequences in terms of eventual customer satisfaction. In general, vehicle lifespan has been increasing, but there should be no expectation that a vehicle will last forever even with good servicing, preventive maintenance, parts replacement, and repairs. This is particularly true for future vehicles with added technical complexity, new system interactions, and higher performance levels. There is a design life for each vehicle and its components, whether expressly stated or implied.

The vehicle lifespan may be more or less than eight, 10, or 12 years. Yet we see vehicles twice that age on the road today. Dealers may offer overaged vehicles for resale as used or preowned vehicles. The vehicles may have been repaired as necessary, but the owner, purchaser, driver, or even the dealer may not know when the vehicle's design life has been exceeded. Such vehicles might be considered 'worn out' either from exceeding the normal lifespan or from accelerated life shortening by abuse, heavy use, or misuse. In any such situation, it is the dealer who is most likely to be in direct contact with the vehicle and its owner. Thus, the dealer occupies the central operative position for dealing with overaged vehicles. It would help if the dealer were specifically advised of the design lifespan. At the expiration time, dealer decisions could be made as to detailed inspections and testing, the installation of replacement parts kits, or other appropriate actions including warnings, technical advisories, and cautions concerning reasonable extensions of the design life. The informed owner would, in effect, knowingly accept any risks from an overaged vehicle.

For future vehicles, it would help if each vehicle's joint service history (from all dealers) could be obtained from an Internet-type system. The system could also suggest special inspection protocols derived from past road and warranty experience, local sites with remanufacture or rebuilding capability, appropriate disposal recommendations, or a buy-back schedule to comply with local environmental requirements. The question might be what are the dealer's choices in terms of the economics for updating or upgrading the vehicle, its dismantling and the value of the parts and accessories, the avoidance of claims for a vehicle in poor condition in continued operation, and the possibility of the sale of a replacement (newer) vehicle? In the future, the vehicles processed by the dealer most likely will exhibit increased technical complexity, diversity of models, and rapid change of specifications. The dealer's choice should not be uninformed, speculative, or beyond reasonably expected capability. Thus, special attention should be given, by others, to risk reduction for the dealer. While it is anticipated that most vehicles will last longer and more will be operated in an overaged condition, the dealers might not anticipate and prepare for the kind of changes and choices that are beginning to occur in a worldwide marketplace. Additional research is needed to determine exactly what is required to enable the dealers to properly support a longer life-cycle for future vehicles in a very competitive marketplace.

(k) Dealer restrictions

The recent reactions of handgun manufacturers to litigation and threats of litigation have resulted in a greater management focus on how to better control the distribution and retail sale of potentially harmful products. This includes imposition of stringent restrictions on sales by dealers. Some gun manufacturers have agreed to ask their dealers to obey specified sales limitations and have redrafted their written contracts to enforce compliance by dealers who carry their brands. Those manufacturers have accepted some responsibility for closer supervision and oversight of their authorized independent dealers.

The word 'independent' provokes strong reactions among dealers, regardless of just how independent they are supposed to be in terms of their franchise or authorization to carry the brand. The gun industry had relied on fully independent retailers; had proven effective legal defenses; slogans that people kill, not guns; a lawful product with deep constitutional overtones; and a history of federal legislative protection. The continued and increasing threat of civil litigation became the menace of corporate death by a thousand bleeding cuts. Similar compelling forces have often changed management policies in other industries and we can expect more to occur in the future.

The future automobile industry may require similar changes in the degree of control of dealers in terms of sales restrictions, training of salespersons, the accuracy of representation to a prospective purchaser, and the cultivation of customer satisfaction. For example, there may be efforts to restrict sales to those with a 'good fit' for a particular brand model vehicle and to convey adequate information to prospective purchasers (particularly safety-related information). This could provoke strong reactions among dealers or salespersons who do not want sales limited in any way or encumbered by a communications burden. However, manufacturer-owned dealerships and consolidated mega-dealers provide an opportunity to develop techniques that might prove effective.

The 'sell anyone as much as possible' attitude among some dealer salespersons may force-fit or coerce customers into future vehicles that exceed their capability or induce accidents. There is a balance between meeting a sales quota and creating a road hazard.

This is not in the manufacturer's best interests. Some vehicle manufacturers are valiantly attempting to improve customer satisfaction at the dealer level and this type of effort is becoming more important to brand loyalty and perceptions of dealer honesty as a service provider. The task will become ever more important in a complex future society. Even though a central register of all vehicles may provide needed historical data for various purposes and permit quick Internet and fax contact with the vehicle owner, the dealer's personal contact with the customer will remain a key ingredient for optimum risk reduction.

(l) Local issues

There are many local issues that could affect the design of future vehicles. For example, where there are many attached or integral garages, should the home be modified or should the vehicles be designed to minimize fuel vapor (evaporative) emissions and exhaust emissions? The concern is with the health effects of volatile organic compounds in general, benzene in particular, and carbon monoxide. A recent study (Mann *et al.*, 2001) indicated that homes with attached or integral garages had benzene levels that grossly exceeded the applicable standards. In one case, the benzene in a room above the garage was 2.5 times the ambient standard. Benzene is a Group 1 carcinogen (International Agency for Research on Cancer). A related issue is the requests by local government agencies to design automobile exhaust systems to be suicide-resistant or to eliminate carbon monoxide emissions. The ambient air pollution issues are also local in character, but up to 70% of the ambient (environmental) benzene emissions have been attributed to road vehicles. What may seem to be a minor constellation of local issues today may become much more in the future and require some form of remedy.

(m) Advanced features in future vehicles

(1) Styling and aerodynamics

Styling changes in future vehicles are inevitable to establish brand identity and demographic sales appeal. The *aerodynamic* concerns, important for fuel efficiency ratings, include swirling and rotating vortices, periodic and unsteady flows, and relatively incompressible airflows. Both styling and aerodynamics directly affect vehicle safety. Sheet metal styling and the attached structures affect crash crush and controlled crush. Aerodynamics, such as smoothing of the underside of vehicles, could be in the form of damage-resistant shields. However, potential future vehicle safety problems may arise from the tendency to make last-minute styling changes and eliminate shielding and skidplates on a cost and simplicity basis. Lightweight aluminum *spaceframes* and body segments with welded stamped metal, extrusions, and cast material have crush and energy management characteristics different from conventional steel frames, structures, and body panels. To prevent problems, this will require considerable testing and time-tested experience, because of the intrinsic variations in the configurations of spaceframes.

(2) Four-wheel steering

Four-wheel steering systems in passenger vehicles and in trucks may improve low-speed maneuverability, but will affect drivability. These systems are generally rear-wheel steering systems intended to supplement the front-wheel steering system. At low speeds, the rear wheels move in a direction opposite to that of the front wheels. But at higher speeds, the rear wheels turn in the same direction as the front wheels. This may improve vehicle stability at higher speeds by reducing vehicle yaw (rotational movement) during lane changing, passing, or evasive actions. It may improve truck stability by reducing corrective steering in wind gusts and trailer sway. At low speeds for cars and trucks, it shortens the turning radius. There may be human problems if the driver can make adjustment to the rear-wheel steering at will, by controls on the instrument panel. Drivability is influenced by degree of steering response, the difference between neutral and active rear-wheel steering, the characteristics of a particular model vehicle combined with its load and attached trailers, the external environmental conditions (ice, snow, rain, wind gusts, and grade) and the sophistication of the driver.

(3) Display integration

There is already a capability of combining or *integrating many displays and controls* into a single liquid crystal display of a large or small size. A single display saves space and the costs of discrete components. But in terms of the human driver, too many functions could add complexity, confusion, and too many choices. What is important enough to display: temperature control, audio control, time, interior lighting, speed, fuel level, the odometer, the estimated time of arrival, hotel information, navigation aids, or other functions and information? What functions should be shared on the display, such as on a split screen, an override, or a permanent overlay? Should an improper physical distance between an occupant and the airbag control unit or instrument panel be displayed until corrected or should the airbag deployment unit itself make an adjustment? If an improper head position is detected should it have display priority? There are many questions and possible design variations which could affect vehicle safety.

(4) Adaptive headlights

Smart headlights, intelligent adaptive front lighting systems, or enhanced lighting systems provide obvious safety benefits. They may swivel by linking steering direction with headlight beam direction, have self-leveling headlights if necessary, may adjust the pattern of headlight illumination on the roadway in terms of beam distance, width, scatter, and symmetry. The European asymmetric 'dipped beam' light pattern, providing illumination toward the outer edge of the road, may be changed to adaptive symmetrical lighting and still meet the regulatory intent by means such as cutoffs that limit the beam pattern. There may be bending lights, either by swiveling (turning) one or more headlights or by increasing the brightness of one of the headlights. There may be weather lights that adjust to fog, rain, snow, or wet roads. There may be city lights that take into consideration the partial illumination that already exists and the low speed of the vehicles in an urban area. There may be highway lighting with a greater beam pattern distance and a cutoff to reduce glare or dazzle to oncoming traffic.

(5) Global warming and emissions

In the early 1990s, *global warming* concerns, particularly the thinning of the ozone layer associated with the chlorine from chlorinated fluorocarbons such as the R12 refrigerant, prompted the switch to the alternative HFC-134a refrigerant. But, even that refrigerant is considered a global warming gas, according to the 1997 Kyoto Protocol. There may be a change to carbon dioxide or some other alternative refrigerant in the near future. The point is that the design of future vehicles will have changes due to public perceptions and environmental beliefs concerning global warming.

Emissions can be reduced by variable engine valve timing, catalysts, hydrocarbon absorbers, and collapsible plastic bladders in the fuel tank. Diesel pollutants, particulates and nitrous oxide may be effectively cleaned by catalytic converters if low-sulfur fuel is used. In the future, fuel cell-powered vehicles may provide the necessary emissions reduction. It has already been demonstrated that propulsion by hybrid-electric systems (gasoline fueled internal combustion engines supplemented by electric motors) provides fuel economy and low emissions benefits. However, it should be recognized that each added component or new refinement may increase the complexity of service and repair operations and the cumulative cost to the vehicle owner.

(6) Design safety research

Research and development is taking place on the following vehicle safety features.

- *Pedestrian safety* may be enhanced by having exterior airbags at the leading edge (over the hood) of the vehicle and in front of the windshield (the cowl area).
- *Rollover protection* may be enhanced by roof crush resistance and tilt angle actuated side airbags that remain inflated for up to six seconds (a longer time taken for rollovers).
- *Four-point safety belts* may reduce upper torso twisting by restraining both shoulders.
- *Child safety* may be enhanced by better emergency trunk release mechanisms and devices to prevent inadvertent closure of the trunk when occupied by children.
- *Integrated child seats* seem to be a necessity since some studies have shown that the portable aftermarket safety seats have up to an 80% chance of being improperly installed or attached to the child.
- Simplification of *cockpit displays* and controls could yield considerable safety benefits and vigorous research continues on this topic.
- Added *hip room* and pelvic supports require further research.
- The *elimination of the B-pillar* and extension of the glass surfaces may affect vehicle crashworthiness.

If the rear seats of a vehicle can be folded, to provide *space for bicycles*, snowboards, and other cargo, does this create a flying object risk for quick vehicle decelerations?

Some *concept vehicles* demonstrate both the creativity of the industry and the possible radical diversity of some future model vehicles.

Constant attention to automotive vehicle safety is needed to prevent regresssion to the unacceptable.

(7) Compromises

Simultaneous engineering may be expanded to include, at the start of vehicle development, a combination of designers, engineers, and marketers who work together at the same location. This is an attempt to achieve *early compromises*, rather than wait for late reviews and costly re-engineering. Late changes may not be fully evaluated, because of scheduling problems, when there are sales-appeal trim items, horizontal spoilers that deploy at high speeds for better handling, the marketing need for a higher horsepower engine, or added interior leg room by other sacrifices.

The design of new vehicles is generally a market-research-driven product development process. Something different in the vehicle may be needed to meet the desires and needs of prospective purchasers, before a competitive vehicle reaches the market and satisfies the consumer demands. The quickest road to travel may be to modify the vehicle body, but keep the platform of an existing model. Another approach is to focus on minor modifications to an existing vehicle within cost and schedule targets.

Not to be overlooked are possible *animated vehicles*, with exterior moving parts, changing color surfaces, blinking headlights, and decorative additions to give the vehicle an appearance of being happy, sad, cute, or responsive to other vehicles. Too many aftermarket novelty modifications might compromise sophisticated vehicle safety features.

(8) Research on conduct

Despite design safety advances, considerable human factors research is needed on popularized hot-button topics concerning *driver conduct*. This includes tailgating, road rage, and high-speed driving. The slogan 'drivers who kill' characterizes automotive vehicles as lethal weapons in the hands of irresponsible drivers. Police pursuits highlight drivers who violate the laws. Are road warriors part of a disorganized army in conflict? In essence, design safety features in future vehicles will benefit many, but driver conduct needs productive research oriented toward error control. But don't blame the driver alone!

(n) Summary

Many new safety devices or features are now being incorporated into the design of automobiles. It is of interest to learn if their benefits are being compromised in any way by the end-user. Studies are being conducted to determine the actual use and the related information provided directly to the purchaser by the dealer. There are issues related to optional safety devices or accessories, out-of-position or misaligned occupants, improper seat adjustment, and purchaser–user misunderstanding of recommended risk-avoidance practices. The objective is to determine if any health promotion efforts might be desirable in the future, by and for dealers, lessors, and vehicle owners.

14 Discussion questions

The following questions may be used for classroom discussions, self-examinations, or reviews of the contents, applications, and scope of practice of the subject matter contained in this book. For privacy concerns, the names and circumstances in the examples have been altered.

1. *John* recently graduated near the top of his class at the state university. He majored in mechanical engineering because he was mentored by a friendly professor who took personal interest in his students. Upon graduation, he was fortunate enough to get a job in the automotive industry. After several months of work, he realized that his work supervisor did not have the time to tutor him, his coworkers were too busy on their projects, and there seemed to be no instructive books available on the job. While he performed the work tasks assigned to him, he was accumulating little of the in-depth knowledge necessary for job promotion or providing added value to his work tasks. He understood that superficial knowledge provided little job security in a cyclic industry. He was particularly concerned about acquiring useful skills in automotive vehicle safety. What should John do? What would be a good short-term and long-term plan to acquire the needed job skills?

2. *Ken* was promoted to the post of research engineer at an automotive parts supplier. He was told to focus on product improvement and cost-savings. The company policy was to reward a good cost-reduction idea with a bonus and personal recognition, if the idea was accepted. In a relatively short time, Ken found a plastic material substitute for one part that could save 54 cents per unit, and the company manufactured over 800,000 of those units annually. The new substitute material met all design specifications and test requirements, so he was elated. He decided not to reveal that there was a tendency for the plastic to crack after three or four years of service, because it occurred only in some environments. Should Ken receive his bonus and move on, or should he reveal the material's potential problem? Should he inquire further, look at test reports, discuss the matter with material specialists, or take other actions?

3. *Mary* purchased a large pickup truck with ample power to haul her horse trailer. After the purchase, the truck did not seem to have sensitive and tight braking on the downside of hills. At one highway exit, she almost rearended another vehicle at a stop sign. When she hitched the horse trailer on the vehicle, it took too long to stop the combination in traffic. She complained several times to the dealer, who made some adjustments, but the brakes still seemed inadequate. Was the dealer at fault for selling the vehicle for a known purpose and failing to correct the problem?

Was one neighbor correct in telling her the vehicle had marginal brakes to start with and it would be best to sell the truck and get a better one? Was this a customer satisfaction issue or was it a disappointed purchaser who complained too much? What are the automobile safety issues?

4. *Paul* was an engineering manager at a well-known supplier of automotive vehicle parts. He was evaluating some engineering drawings as part of an invitation to bid on a large contract. He noticed some specified materials had the words 'or equivalent' after the description of the materials. He realized that some, possibly equivalent, substitute materials might be available at lower cost and that any cost savings could increase the company's chances of winning the large contract. He telephoned some salespersons to see what was available, at what cost and terms, and possible delivery schedules. He then approved a contract bid that pleased the procurement people at the vehicle assembler, who were under orders to effect a 15% reduction in parts costs if at all possible. They got the contract, then a sense of reality set in, and they wondered whether there would be any trouble with the part after it was manufactured. What should Paul have done to assure acceptable quality and safety of the part?

5. *Mike* was a marketing specialist in a rapidly growing vehicle accessory manufacturing firm. He regularly visited some dealers, distributors, and retailers. His prime job and main focus was to promote sales, but he observed some situations involving his products that seemed suspect as to safety. Although at first he dismissed his concerns because he was not a safety specialist, he gradually came to the belief that he had to protect his company's image if there were to be any increases in future sales and sufficient purchaser trust in the integrity of the product. What could Mike do without directly challenging his customers? Should he leave product safety in the hands of the design specialists?

6. *Bill* conceived an important improvement in the design of a subassembly. He had labored hard and long for the victory over a marginal and problematic design. The new design was accepted, but recommendations were made to provide test points and to assure that it could be manufactured at a reasonable cost. Bill indicated that building-in test points and determining manufacturability was not his job. Was he correct? How could these recommendations be accomplished? Assuming others take those responsibilities, what 'engineering actions' by Bill are still necessary?

7. *Jim* was given the task of 'man testing' some construction vehicles. There had been complaints, on earlier models, that access to different locations on the machine was difficult and precarious during the operation of the machine. How could Jim develop a good plan for usability testing and of what would it consist? What design criteria would he utilize? Would he use human subjects similar to construction workers?

8. *Barbara* was employed by a major vehicle assembler as a design engineer. She coordinated the design of electrical harness bundles with a supplier. The supplier was assigned considerable design, manufacture, test, and quality responsibilities. Several months after production of the vehicle started, there were field complaints about damaged electrical harnesses. What should Barbara do to find out where and how the damage was occurring? What could she do to prevent the damage?

9. *June* purchased a rugged appearing SUV, advertised as having off-road driving capabilities, including four-wheel drive. After a few weeks, she decided to test the vehicle by going off the road, cross-country, to visit some friends in a hilly rural area.

She hit some rocks and damaged something under the vehicle, went further and got stuck in a muddy stream, then abandoned the vehicle and walked to find help. She loudly criticized the vehicle to her friends. Was she fair and correct in her criticisms? What constitutes appropriate off-road use? What design features should be built-into a vehicle that is advertised for off-road use? Should people believe advertising and sales promotion pitches? Are there design safety considerations in this scenario?

10. *Harry* purchased a children's motorcycle for his four-year-old son. It was powered by two 6-volt batteries, had a hand control accelerator, a foot brake pedal, electric lights, electric horn, dual rear-view mirrors, and training wheels. The cost was about $300 (US), but Harry also provided a leather jacket and dark sunglasses for a 'good look' for his son. The look and feel were very real for the future motor-cyclist, who was very excited about the gift. As neighborhood parents, what is your reaction? What could Harry do, for safe use, and still please his child? What are the appropriate and adequate controls, by the manufacturer and local government entities, to minimize the risks of injury in this situation?

11. *Martha* was designated manufacturing representative, instructed to assure good coordination with the engineering department, during early design and development of a new farm equipment vehicle. During the initial formal design review session, she attempted to discuss the use of existing production machinery and processes in order to reduce manufacturing costs. The design engineers were enthusiastic about their new design and did not want to modify it just to please manufacturing. What should Martha do to accomplish her assigned tasks?

12. *Jonathan* was a quality control engineer, a black belt in achieving zero defects. He worked assiduously on incentive and motivation programs. He also implemented quality inspection, by interpreting engineering, production, and assembly drawings and other documents. This seemed to generate an endless list of quality requirements that were far too complicated and costly, so something needed to be done to eliminate most inspections. What should Jonathan do? What inspections are actually important to the company? How does he determine what should be eliminated, what is cosmetic, and what should be considered safety-critical?

13. *In Europe*, there is a pallet truck of 6-ton capacity that is battery powered, rider operated, and can travel at speeds up to 3.1 mph (5 km/h). It has 'finger-sensitive electronic proportional steering' as a design safety feature. Pallet trucks have been a significant source of injury and damage in material handling and warehouse operations because of the maneuverability required of such vehicles in limited factory spaces and narrow aisles, and among inventory stacks and workstations. For the operator, why would proportional steering be a benefit? Why should the steering be finger-actuated? What other design safety features would be desirable? Would forklift gloves (urethane bonded to the fork tines or tips) reduce injuries and damage.)

14. *In England*, when lifting heavy loads near the limit of a crane boom's reach, there had been a series of overturning accidents. In response, the Supply of Machinery (Safety) Regulations 1992 required interlocks to prevent operation of truck-mounted cranes (lorry loaders) without deployment of their stabilizers. However, this regulation only applied to new vehicles and contained the term 'or equivalent engineering solutions'. Optional interlock kits became available, but overturning accidents continued. Is the danger so obvious that government regulation should yield to better trained and more careful operators? Would tipover alarms be a better

choice than interlocked stabilizers? Are exceptions, in government regulations, always necessary?

15. *In some countries*, the flat rooftops of factory buildings are often used as patio areas or for worker rest and relaxation purposes. They are also used for maintenance access to rooftop air-conditioning units, roof lights, and electrical equipment. This use is opposed by many architects as destructive to the roof membranes and leak-producing. What type of roof edge protection is needed to prevent workers from falling off the roof? Should such roof tops be used by lookout coordinators to visually spot, identify, and communicate with trucks carrying parts to the factory and those departing with vehicle subassemblies destined for vehicle assemblers under a just-in-time delivery schedule?

16. *Sarah* experienced troubles with the navigator in her new passenger sedan. As she traveled down the highway, she had to bend to the right to see the correct control buttons and the display map seemed to scroll in the wrong direction. A little later it told her to turn right, but there was no exit for another 5 miles. When she looked again, her vehicle position was located in a nearby river. About that time she stopped on the shoulder to determine what was wrong with the navigator, even though stopping on the shoulder was a violation of the traffic laws. Was this a typical or atypical safety distraction problem? Why did the navigator operate while the vehicle was traveling down the highway? Should the navigator have had controls and displays that were more human-centric and better oriented so that head movement was minimal?

17. *Harry Philo*, a Past President of the American Association of Trial Lawyers, crusaded against surface-level railroad-grade crossings. He frequently asserted that there could be no accidents or collision deaths if the roadway was separated from the railroad tracks by a bridge. If this is true, why have dangerous crossings persisted? How could the train–auto risks be minimized? The bullet trains of *Japan* have enviable safety records with separated grade crossings and good signal devices. In the *United States*, about 80% of grade crossings have no safety protective features. For train approaches see *Railroad–Highway Grade Crossing Protection – Recommended Practices*, Bulletin 6, Association of American Railroads. Why are the lessons learned from one culture not transferrable to another, particularly if they could save lives?

18. In the *United Kingdom*, after the Ladbroke Grove rail crash (1999), the accident investigation caused Lord Cullen (2001) to recommend a strong Rail Industry Safety Body, more precise government regulations, advanced signal systems, and improved train crashworthiness. What effect can be anticipated from high-level government accident investigations and recommendations? How could railroad trains be designed for improved crashworthiness of engines, passenger sections, and cargo carriers? Are the basic safety principles and concepts that are applicable to on-road passenger vehicles also applicable to on-rail passenger trains?

19. *Survivability* in vehicle crashes has become a major design objective. It includes escapeworthiness from tangled crushed metal and fuel-fed fires. In *England*, the Railway Safety (Miscellaneous Amendments) Regulations 2001 place great importance on the escape from trains in emergencies as well as the organized evacuation of trains. Is escapeworthiness compatible with crashworthiness? How can a fire-risk assessment be performed? How can telematics affect survivability?

20. A *bicyclist* observed about six vehicles on an off-ramp that were stopped for a red light. He started across the intersection in front of three abreast vehicles. As he passed the first vehicle, an obviously inattentive driver quickly accelerated to make a permissible right turn. The left front of the moving vehicle impacted the bicyclist, he flew over the corner of the vehicle, and landed on the roadway surface to the left of the vehicle. He stood up and claimed no injuries, but bloody scrapes could be seen on his right forearm. The driver of the impacting vehicle quickly apologized. Both the bicyclist and the driver refused to swap identification and both departed the scene because of the heavy traffic. Are all accident injuries immediately known to the participants? Could the conspicuousness of the bicycle have been a factor in the accident? Was the lack of serious injury due to the smooth front end of the vehicle, i.e. no projections and no sharp angles in the sheet metal?

21. *Sam* reads the owner's manual only when he has trouble with his automobile. When he attempted to read it when a question arose, he found it difficult to understand. He did have a minor literacy problem in the English language because his primary language was acquired in a foreign country. How can functional illiteracy be overcome for automobile drivers? Should vehicle publications be measured for readability and, if so, what grade level or reading ease score would be acceptable? How are drivers with different languages and literacy skill levels accommodated in the *European Union*, where laws dictate that there be no impediment to travel among that community of nations? What might happen if Sam received a written vehicle recall notice in the mail?

22. A *fleet manager* claims to have 'seen it all' from speeding tickets to crazy collisions, from sleepy drivers to drunken and drugged drivers, from those who are dependable to those who feel entitled to a small part of every load they carry (damaged goods), and from straight-line deliveries to unnecessary detours. While he tries to 'cover' for the drivers' mistakes and intentional deviations, in order to stay in business, he also plays the 'blame game', placing fault on others. When asked about accident statistics, he believed they reflected only excuses and cover stories. How could the fleet manager correct problems, assess trends, or make predictions if his database was concocted and unreliable, and full of errors as to causation? What could be done to achieve policies and procedures that could provide useful information?

23. *Tim* was surprised at the number of cupholders in his new car. He rose to the challenge, brought some beverages, then drank with one hand holding the cup and the other on the steering wheel. He remembered that one-hand driving had been prohibited with the outlawing of spinner knobs on the steering wheel. But he liked the idea of stopping on the way to work to get a cup of coffee and some doughnuts to eat during the morning traffic crunch. Is it true that automobile manufacturers, by supplying cupholders in their vehicles, were encouraging one-hand driving? If the cupholders were of the retractable type (out of sight, out of mind), could there be any inference of encouraging one-hand driving? Would the vehicle occupants drink and eat in a vehicle even if no cupholders were provided? Are cupholders a safety problem or only a convenience item?

24. A company with a large 'campus' had what it considered to be too many workplace accidents involving transport vehicles. A consultant advised it to develop a way of separating pedestrians from the vehicles. In some locations, he suggested segregated traffic routes for guest (visitor) automobiles, delivery trucks, interbuilding forklift trucks, personnel vehicle shuttles, and roads to unloading and storage areas. He also

recommended better training, testing, and the use of interactive computers for refresher training, data collection, and vehicle dispatch operations. Does this advice sound like costly overkill? What would be more important: traffic separation, driver training, interactive computers, improved dispatch operations, or a different selection of vehicles?

25. An automotive vehicle manufacturer had some engineers who specialized in failure modes and effects analysis (FMEA), a design tool to identify potential failures, causes, consequences, and prevention. Much of their success was due to their effective coordination with other organizational departments in order to generate logic diagrams, estimate quantitative risks, and secure design improvements. Top management decided to extend this design tool to the design transition to manufacturing, a process evaluation technique (PFMEA) that might eliminate variations from the design intent and achieve earlier product consistency. How does the PFMEA differ from conventional quality control? Is there any benefit in applying such detailed techniques to the production process? Is this an attempt, by design, to impose strict process-specific rules?

26. After a product is released to manufacturing, there is a concern that the product, as produced, may not be exactly the same as the design intent or that the expected variations may be too great for a 'quality product'. There are many quality standards that provide guidance for statistical sampling, such as those of the ISO, ANSI, ASQC, and DOD (USA). Most companies have some kind of an inspection sampling plan, for attributes and variables, to demonstrate conformance of the product to the contract and technical requirements. Under some standards, a lot may be accepted with one non-conforming item and, if the attribute is safety-related, there may be some acceptable consumer risk. This contrasts with the high cost of 100% inspection (rather than sampling) or the rejection of every lot with a non-conformance (a perfection approach). Should a non-conforming part be considered a defect? If the risk is small, should the lot be accepted? How should the quality assurance department be advised of those attributes and variables that should be considered safety-critical?

27. There is a rule that defects can be inserted into a product anywhere. Is this true? If so, what can be done to minimize such defects? Could participatory intervention techniques have any effect on the rate of insertion?

28. There are those who believe that the 'instinct' of the designer, his 'gut feeling', and his 'free-spirited' creativity should be allowed full uninhibited expression. After a model is constructed, it is shown in clinics to determine customer appeal and management acceptability. Concept vehicles appear at auto shows to get the reaction of a broader base of potential customers, peers, and media critics. Is this process of creation, clinics, and shows the way to go for vehicle safety? What kind of market research could affect the introduction or deletion of design safety features?

29. Remarketers (used-car salesmen) may purchase repaired vehicles at auction, then sell them as previously owned vehicles. Some unrepairable vehicles have been sent to salvage yards, only to reappear for sale to the general public. Flood-damaged vehicles may be sold to other dealers for resale as new vehicles. Some luxury vehicles, with limited mileage by a prior owner, are sold as new cars. If the vehicles are in good condition, should the final purchaser complain? If there is a monetary discount and an oral disclosure that the vehicle is 'like new', is this sufficient? Are these practices something of a safety problem? What kinds of disclosure would be appropriate?

30. Some accident reconstructionists have gathered data regarding pedestrian impacts. They claim that pedestrians hit while walking away from the vehicle will tend to have their body sit on the hood, slide towards the windshield, then fall off. If struck facing the vehicle, their bodies may wrap around or vault the vehicle. With a fender vault they may strike the ground or roadway to one side of the vehicle, there may be a head strike on the windshield. If the vehicle is decelerating, there may be a somersault and a landing in front of the vehicle. How does vehicle speed change the location of a pedestrian head-strike with a roof vault? Is head injury affected if the head snaps or whips against the vehicle body? If a van or bus with a relatively high flat-face vehicle front strikes a pedestrian, what biokinetic motion occurs? How can vehicle design reduce such pedestrian injuries? How reliable are the available data applied to varying circumstances?

31. *Helen* was asked by her design supervisor to 'research' a recommendation that a new model vehicle should not have carpeting on the floor. Some stylists (designers) believed they could achieve a 'new look' by having bare floors. The marketing appeal was to the demographically young or entrance-level purchaser who was rebellious and wanted his vehicle to be different. The floor could be coated in various colors and designs. In terms of human factors, what would be the effect in terms of vehicle interior noise, vibration, access, and other factors? If there were human factor problems that were not safety-related, should this affect those who wanted such a vehicle to be marketed?

32. *Pat*, having heard about the driver distraction problem and seeing so many drivers holding a cell phone to their ear as they made sharp turns, decided to do something. He purchased a hands-free in-car telephone mount so he could keep both hands on the steering wheel during telephone conversations. He ordered a voice command interface device. He asked about a voice recognition system for interface simplification (no strangers making calls and no buttons to look for and depress), so his eyes could stay on the road. He did not want to pull over to the side of the road just to make or receive a telephone call. Were Pat's responses appropriate for the risk? Which alternatives were both reasonable and effective? Would the mental work load, in heavy traffic, trigger aggressiveness or road rage?

33. *Jill* was concerned about her safety when driving through a crime-ridden area known for hijacking or commandeering of vehicles at traffic intersections and stops. She decided she needed a panic alarm for an immediate response to anyone or any condition that might threaten her. What type of panic alarm would be most appropriate? Should it be an OEM device or an aftermarket device? Would there be increased risk as well as a search and rescue benefit? Was her fear justifiable?

34. *Spence* attempted to climb into the cab of his SUV. It was raining and the running board was wet. The plastic covering of the running board had smooth ridges oriented in alignment with the front to rear of the vehicle. The ridges seemed to provide drainage channels and slip-resistance for the foot in the direction of the movement into the cab. The leather sole of his right foot contacted the plastic and he shifted his weight on to his right foot as he climbed upward. Then, his foot glided over the smooth ridges and his leg moved sideways to his right. He lost his balance and his grip on the B-pillar, then fell to the ground. Could this safety problem have been corrected by a block pattern that provided slip-resistance in two directions? Did he have a good handhold? Was he careless?

35. There are advanced design technologies such as fingerprint identification pads within and outside the vehicle to deter car theft and eliminate car keys. There are devices for automatic headlight leveling (aiming) in accordance with the load carried on the suspension system. There are and will be a variety of impressive telematics devices. Each has an added cost. Some involve a usage fee or a monthly subscription charge. What is the acceptable cost for each such benefit? How much of a price increase for a vehicle is tolerable, if it is well equipped with such devices?

36. Children have equal or better visual acuity than adults, yet are more likely to be involved in pedestrian accidents where vision is a contributing causal factor. Children may estimate the speed for an approaching vehicle by its change in size against a background. Speed determination may be easier when the vehicle crosses a field of view. Do children use simplistic assumptions before crossing a roadway, such as seeing the rear of a vehicle and assuming it is traveling away from the crossing? Is it a matter of making a combined judgment of size, speed, and distance? Is it a matter of selecting the wrong location to cross the road? Is it a matter of road-crossing training? Are there any vehicle design safety changes that might help prevent child road-crossing accidents or minimize the child's injuries?

37. Jackknifing of tractor trailer rigs may be caused by incorrect braking (locking-up the rear wheels of the tractor). But what causes trailer swing? What effect does load shifting have on jackknifing, trailer swing, or tipover?

38. The aggressive driver has been alleged to be a significant cause of vehicle accidents. There are those who believe that automobile drivers should be free to be independent and express their feelings. There are those who believe that they are a menace to others. There are those who believe that aggressiveness is acceptable, but that traffic laws should also be aggressively enforced. Still others believe that automotive vehicles should not have excessive horsepower, acceleration capability, steering without seat-of-the-pants feeling, and advertising implying that driving is a competitive sport. What balance might be achieved among these conflicting beliefs or opinions? Who could effect the greatest change in driver behavior? Should any effort be made to control aggressive driving? Are there any reasonable design safety measures that might be taken?

39. A dealer service advisor telephoned a customer to tell him that a leak had been found in the oilpan during the scheduled servicing of his vehicle. When the customer asked what caused the leak he was told there was a hole, probably caused by an impact, and that he must have run over something. The cost of parts and labor would be about $765 (US). When he asked if the hole could be plugged, the service advisor replied 'it's made out of aluminum'. The repair was approved by the customer. When he went to pay the bill, the invoice read 'oilpan broken'. He asked for the old oilpan, examined it and found that it was not broken and had no holes. He leak-tested it and found that it did not leak. Why would a dealer service advisor want to replace an oilpan rather than just a gasket? Should the dealer be interested in customer trust and good service in order to sell new vehicles? Is dealer reputation important or would this be communicated to only a few prospective customers? Would it affect the manufacturer's brand image?

40. The 'give-room' behavior of a motorist on a two-lane highway may be illustrative of human speed estimation difficulties. Are most vehicle drivers overconfident of their ability to estimate the speed of other vehicles in terms of 'headway distance'

or 'gap size', their 'neural feedback' or 'motion analysis', and their capability in terms of 'evasion maneuvers'? What design safety devices might aid the driver in this respect?

41. A memo from an independent test laboratory used the term 'negative transfer errors' in regard to an all-electronic brake system for a new model passenger vehicle. What was meant? How could this be tested? Does it represent a possible latent safety defect?

42. There may be some risk, in the future, of hackers who attempt to gain entrance into an automobile telematic system. This might happen when there are remote wireless systems for providing the driver with vehicle maintenance and diagnostic information. It could occur through vehicle tracking systems. Would hackers attempt to destroy safety-related systems? What might constitute an acceptable level of security?

43. Automobile assembly facilities have been constructed that move the vehicle bodies on skillets that can be raised, rotated, and lowered for worker accessibility. Robotic or automatic welders, online lasers for the inspection of the accuracy of build, and the immediate proximity of production engineers are key features of such plants. There is more open space and better lighting. These facilities have been called 'white factories' as compared with the older 'dark factories'. The question is which type of factory is most cost-competitive? Should there be a difference in product discrepancies and defects?

44. The stopping distance of a vehicle is dependent on the braking system and how fast the driver actuates the brakes. Conventional hydraulic brakes have a master cylinder and hydraulically actuated calipers at each wheel. With a vacuum-assisted power booster, the brake system can be actuated in about 300 ms. With a proposed design of an electrohydraulic (an electrically driven motor pump) reservoir of high-pressure brake fluid and an electrically controlled valve release, the brake actuation time can be reduced to 100 ms. At 60 mph (88 ft/sec) this could reduce the vehicle stopping distance by nearly 18 feet. Since this retains the driver in-the-loop, it is still human-dependent and prone to human error. A computer determines pedal pressure and its speed of application, in order to decide how much brake pressure should be applied. The fail-safe feature is that the brake pedal can still actuate the brakes if there is an electrical or computer failure. Is the stopping distance reduction worth the added complexity of a high-pressure reservoir of brake fluid, the electrical valves, and computer control? What is the risk of being rear-ended during a fast stop?

45. To achieve full advantage of electronic accident avoidance systems, a full brake-by-wire (electrical brake actuation system) probably will be necessary. The driver's response is just too slow, in terms of human capability, compared to radar object detection, computer analysis, and signals to the braking system. If there are no mechanical or hydraulic connections between the brake pedal and the brakes at the wheel (the driver is out of the loop), does this create safety problems? If there were a combination of braking, suspension, and steering responses that were all computer-controlled and coordinated, would that make a difference? Should the system be fail-safe and fault-tolerant (redundant)?

46. A new product development organizational structure for a vehicle assembler was developed to reduce the time from design sketches to working models, so they could be shown to the public for reaction early rather than later. The structure consisted of product planning and competitiveness, styling and design, engineering development and testing, purchasing and manufacturing, and marketing and sales.

Could they all work together in a harmonious manner? Could this reduce product development time and cost? At what points should vehicle safety play a major role? What is the role of focus groups in product schedule compression?

47. An active roll control system was being considered to eliminate rollover potential. The basic theme was that sensors detect and actuation offsets the undesirable. This was applied to stabilizer bars to reduce body roll. Active suspension was considered. A full-scale system involving many factors was discussed. Just how important is roll control? How many subsystems should be integrated in a vehicle stability and rollover prevention design effort?

48. As the driver increases the speed of his vehicle, its kinetic energy increases. Kinetic energy due to the motion of the automobile is the capacity to do work (cause destruction and change the velocity in other objects). Accident reconstructionists often recite the principle of conservation of energy: that energy can neither be made nor destroyed, it can only be changed in form. If kinetic energy is expressed by the equation $KE = \frac{1}{2}mv^2$, how would you characterize the gain in the vehicle's capacity to cause destruction as the speed increases? What is the difference between the crushing ability of a vehicle with an impact speed of 30 mph as contrasted with 60 mph? How would this energy be expended against the resisting force of another vehicle when it is impacted?

49. The body of a passenger vehicle is generally composed of relatively inexpensive low-carbon steel sheets and formed structures. The steel is capable of deep drawing or formability and there is an ease of resistance welding. However, the quality of the weldments depends on welding current, current time, electrode force, and electrode characteristics. For example, increasing the welding time increases the shearing strength and the button diameter. While resistance welding requires little skill, there has been a shift to automatic welding machines (robots) for greater uniformity and cost reduction. The process variations do require diligent welding inspection procedures. Are human errors to be expected during set-up of the steel to be welded, in the welding process itself, and in the inspection of welds? If so, how can the human errors be minimized?

50. Automotive vehicle engines for small passenger vehicles are designed in three primary versions: (1) the European engine that cruises smoothly at high speeds, (2) the Asian engine that is very economical in terms of fuel consumption, and (3) the United States version that provides quick acceleration. Is this an example of marketing theory dictating design? Does this reflect cultural differences, such as driver aggressiveness? Would a true world-class vehicle, with one engine, be more appropriate?

Notice: If these discussion questions induced some differences of opinion and uncertainty, this reflects the experiences of those in the automotive industry who must make choices quickly to meet pressing schedules.

15 References and recommended reading

AASHTO, 1973, *A policy on design of urban highways and arterial streets.* Washington, DC: American Association of State Highway and Transportation Officials.

AASHTO, 1976, *Maintenance manual.* Washington, DC: American Association of State Highway and Transportation Officials.

Accident reconstruction: analysis, simulation, and visualization, SP1491. Warrendale, PA: Society of Automotive Engineers.

Airbag Technology, 1999, SP1411. Warrendale, PA: Society of Automotive Engineers.

Airbag Technology, 2001, SP1615. Warrendale, PA: Society of Automotive Engineers.

Alexander, G.J. and Lunenfeld, H., 1999, Positive guidance. In Peters, G.A. and Peters, B.J., *Warnings, Instructions, and Technical Communications.* Tucson, AZ: Lawyers and Judges Publishing Co.

Alm, H. and Nilsson, L., 1994, Changes in driver behaviour as a function of handsfree mobile telephones. *Accident Analysis and Prevention*, 26, 441–451.

Amstadter, B.L., 1971, *Reliability mathematics.* New York: McGraw-Hill.

ANSI A14.5, 1974, *American National Standard Safety Requirements for Portable Reinforced Plastic Ladders*, pp. 7, 17. New York.

ANSI Z535.4, 1998, *Product Safety Signs and Labels. American National Standard.* Rosslyn, VA: National Electrical Manufacturers Association.

AS9001A, *Quality Management Systems Aerospace Requirements.* Based on ISO 9001:2000. Also see AS9139, *Nonconformance Documentation.*

Automotive Terminology: English–German–French, 1998. Warrendale, PA: Society of Automotive Engineers (Dist.) and Robert Bosch GmbH.

Baerwald, J.E. (ed.), 1976, *Transportation and Traffic Engineering Handbook.* Englewood Cliffs, NJ: Prentice Hall.

Baker, J.S., 1976, Traffic accident analysis. In Baerwald, J.E. (ed.), *Transportation and Traffic Engineering Handbook*, Institute of Traffic Engineers. Englewood Cliffs, NJ: Prentice Hall.

Begeman, P.C. and King, A.I., 1975, The effect of variable load energy absorbers on the biodynamic response of cadavers, paper 751168. Warrendale, PA: Society of Automotive Engineers.

Black, H.C., 1968, *Black's Law Dictionary* (4th edn.) St Paul, MN: West Publishing Co.

Black Nembhard, H., 2001, Controlling change: process monitoring and adjustment during transition periods, *Quality Engineering* 14(2), 211–244.

Breed, D.S. and Castelli, V., 1991, Design of air bag systems. In Peters, G.A. and Peters, B.J. (eds), *Automotive Engineering and Litigation*, vol. 4, pp. 413–458. New York: Wiley.

Briem, V. and Hedman, L.R., 1995, Behavioral effects of mobile telephone use during simulated driving. *Ergonomics*, 38, 12.

Brookhuis, K.A., De Vries, G. and De Waard, D., 1991, The effects of mobile telephoning on driving performance. *Accident Analysis and Prevention*, 23(4), 309–316.

Brown, I.D., Tickner, A.H. and Simmonds, D.C.V., 1969, Interference between concurrent tasks of driving and telephoning. *Journal of Applied Psychology*, 53(5), 419–424.

Brown, J.F. and Obenski, K.S., 1988, Motorcycle accident reconstruction. In Peters, G.A. and Peters, B.J. (eds), *Automotive Engineering and Litigation*, vol. 2. New York and London: Garland Publishing.

Cases from the Special Crash Investigation Program, 1999. Washington, DC: National Highway Traffic Safety Administration.

Cesarani, A., Alpiniini, D., Boniverver, R., Claussensen, C., Gageygey, P., Magnussonson, L. and Odkvist, L. (eds), 1996, *Whiplash Injuries: Diagnosis and Treatment*. Warrendale, PA: Society of Automotive Engineers (Dist.) and Springer-Verlag.

Chaffin, D.B., 2001, *Digital Human Modeling for Vehicle and Workplace Design*. Warrendale, PA: Society of Automotive Engineers.

Chan, C.-Y., 2000, *Fundamentals of Crash Sensing in Automobile Air Bag Systems*. Warrendale, PA: Society of Automotive Engineers.

Chemical Manufacturers Association, 1985, *Risk Analysis in the Chemical Industry*. Rockville, MD: Government Institute.

COM, 1990, *General Product Safety*. See 'Commission Amended Proposal for a Council Directive concerning General Product Safety', COM (90) 259 final-SYN 192, 90/C 156/07. Submitted by the Commission on 11 June 1990. *Official Journal of the European Communities*, No. C156/8014, 27 May 1990.

De Leo, R., Cerri, G., Claretti, L., Primiani, M., Moglie, F., Moscariello, M. and De Riso, M., 2000, Cars: modeling the electromagnetic field for radiated immunity tests. *Compliance Engineering*, Mar./Apr., 36–45.

Driving-Safety Systems (2nd edn),1999, Warrendale, PA: Society of Automotive Engineers (Dist.) and Robert Bosch GmbH.

Drory, A., 1985, Effects of rest and secondary task on simulated truck-driving task performance. *Human Factors*, 2(2), 201–207.

EA 00-023, NHTSA, DOT, USA, 10/8/01.

ECPD, 1974, Engineers shall hold paramount the safety, health, and welfare of the public in the performance of their professional duties. Section 1, *Suggested Guidelines for Use with the Fundamental Canons of Ethics*, Engineers' Council for Professional Development, New York. Also in the Fundamental Canons, *Code of Ethics of Engineers*, ECRD, 1974.

EEC, 1989, Council Directive of 14 June 1989, on the approximation of laws of the Member States relating to machinery 89/392/EEC. *Official Journal of the European Communities*, No. L183/9–14, 29 June 1989 (also see Annex I to VII, *OJEC*, No. L183/15.32, 29 June 1989).

Evans, F.G., 1973, *Mechanical Properties of Bone*. Springfield, IL: Charles C. Thomas.

Ewing, C.L., Thomas, D.J., Lustick, L., Becker, E., Willems, G. and Muzzy, W.H., 1975, *The Effect of the Initial Position of the Head and Neck on the Dynamic Response of the Human Head and Neck to -Gx Impact Acceleration*, paper 751157. Warrendale, PA: Society of Automotive Engineers.

Fairclough, S.H., Ashby, M.C., Ross, T. and Parkes, A.M., 1991, Effects of handsfree telephone use on driving behavior. *Proceedings of the ISATA Conference*, Florence, Italy.

Felicella, D.J., 2001, Skateboard and scooter acceleration trials. *Accident Investigation Quarterly*, Winter, 19 and 48.

Foster, K.R., Vecchia, P. and Repacholi, M.H., 2000, Science and the precautionary principle. *Science*, 288, 12 May, 979–981.

Frank, R., 1993, Automative safety through microelectronics. In Peters, G.A. and Peters, B.J. (eds), *Automotive Engineering and Litigation*, vol. 5. New York: Wiley.

Goerth, C.R. and Peters, G.A., 2000, Product hazard communications. In Stern, W., *Packaging Forensics*, pp. 43–50. Tucson, AZ: L & J Publishing Co.

Gurdjian, M.D., 1970, *Impact Injury and Crash Protection*. Springfield, IL: Charles C. Thomas.

Hancock, P.A., Simmons, L., Hashemi, H., Howarth, H. and Ranney, T., 1999, The effects of in-vehicle distraction on driver response during a crucial driving maneuver, *Transportation Human Factors*, 1(4), 295–309.

Health and Safety Executive, 2001, *Reducing Risks, Protecting People* (www.hse.gov.uk/dst/r2p2.pdf). London: HSE.

Hendler, E., O'Rourke, J., Schulman, M., Katzeff, M., Domzalaski, L. and Rodgers, S., 1974, *Effect of Head and Body Position and Muscular Tensing on Response to Impact*, paper 741184. Warrendale, PA: Society of Automotive Engineers.

Hendrick, H.W. and Kleiner, B.M, 2001, *Macroergonomics: an Introduction to Work Systems Design*. Santa Monica, CA: Human Factors and Ergonomics Society.

Henson, R., Shallice, T. and Dolan, R., 2000, Neuroimaging evidence for dissociable forms of repetition priming. *Science*, 287, 18 Feb., 1269–1272.

Holt, D.J., 2000, *42 Volt Systems: Technology Implications for the Automotive Industry*. Warrendale, PA: Society of Automotive Engineers.

HS (G) 48, *Reducing Error and Influencing Behavior*. London: Health and Safety Executive.

Hulbert, S., 1976, Driver and pedestrian characteristics. In Baerwald, J.E. (ed.), *Transportation and Traffic Engineering Handbook*, pp. 38–72. Washington, DC: Institute of Traffic Engineers.

Hulbert, S., 1984, The driving task. In Peters, G.A. and Peters, B.J. (eds), *Automotive Engineering and Litigation*, vol. 1, pp. 359–382. New York and London: Garland.

Human Tolerance to Impact Conditions Related to Motor Vehicle Design, SAE Information Report J885a, revision of Oct. 1966. Warrendale, PA: Society of Automotive Engineers.

Hunter, A.G.M., 1993, Review of research into machine stability on slopes. *Safety Science*, 16, 325–329.

Hyde, A.S., 1992, *Crash Injuries: How and Why They Happen*. Warrendale, PA: Society of Automotive Engineers (Dist.) and Hyde Association.

ITIIC, 1994, *Explanation of PL Law*. International Trade and Industry Inspection Committee, Consumer Economic Section, Industrial Policy Bureau, Ministry of International Trade and Industry, Tokyo, Japan.

JNPA, 1996, Study of injury producing crashes during June, 1996, Japanese National Policy Agency. *Daily Automobile*, 17 Aug. Also *Wireless Week*, 24 Mar. 1997, 1.

Johansson, G. and Rumar, K., 1971. Drivers' brake reaction times. *Human Factors*, 13(1), 23–27.

Johnson-Laird, P.N., Legrenzi, P., Girotto, V. and Legrenzi, M.S., 2000, Illusions in reasoning about consistency. *Science*, 288, 21 Apr., 531–532.

Kames, A.J., 1978, A study of the effects of mobile telephone use and control unit design on driving performance. *IEEE Transactions on Vehicular Technology*, VT-27(4), 282–287.

Karnes, E.W., Leonard, S.D. and Johnson, F.H., 1990, ATVs: human factors and engineering evaluations. In Peters, G.A. and Peters B.J. (eds) *Automotive Engineering and Litigation*, Vol. 3. New York and London: Garland Publishing.

Karwowski, W., Genaidy, A.M. and Ansour, S.S. (eds), 1990, *Computer-aided Ergonomics*. London: Taylor & Francis, 138–156.

Kastner, S., Deweerd, P., Desimone, R. and Ungerleider, L.G., 1998, Mechanisms of directed attention in the human extrastriate cortex as revealed by function MRI. *Science*, 282, 2 Oct., 108–111.

Lundstrom, L., Kelly, C., Alonzo H. and La Belle, D.J., 1964, *Crash Research for Vehicle Safety*, Society of Automotive Engineers paper 640186. Also in *Highway Vehicle Safety*, 1968, pp. 93–105. Warrendale, PA: Society of Automotive Engineers.

Lundstrom, L.C., 1967, The safety factor in automobile design. SAE paper 660539 in *SAE Transactions*, vol. 75.

Maclure, M. and Mittleman, M.A., 1997, Editorial: Cautions about car telephones and collisions. *New England Journal of Medicine*, 336(7), 501–502.

MacCollum, D.V., 1988, Rollover protective systems (ROPS). In Peters, G.A. and Peters, B.J. (eds), *Automotive Engineering and Litigation*, vol. 2. New York and London: Garland Publishing.

McFarland, R.A. and Moore, R.C., 1970, *Ergonomics and Motor Vehicle Safety*, paper 700362. Warrendale, PA: Society of Automotive Engineers. Also published in *International Automobile Safety Conference Compendium*, 1970, pp. 133–152. Warrendale, PA: Society of Automotive Engineers.

McHenry, R.R. and Nalb, K.N., 1966, *Computer Simulation of a Crash Victim – a Validation Study*, paper 660792, p. 2.

MacKay, G.M., 1973, *The Effectiveness of Vehicle Safety Changes in Accidental Injury Reduction*, paper C194/73. London: Institution of Mechanical Engineers.

McKnight, A.J. and McKnight, A.S., 1991, *The Effect of Cellular Phone Use upon Driver Attention*. Washington DC: National Public Services Research Institute.

McKnight, A.J. and McKnight, A.S., 1993, The effect of cellular phone use upon driver attention. *Accident Analysis & Prevention*, 25(3), 259–265.

Mann, H.S., Crump, D. and Brown, V., 2001, Personal exposure to benzene and the influence of attached and integral garages. *Journal of the Royal Society for the Promotion of Health*, 121(1), 38–46.

Manual on Uniform Traffic Control Devices, 1988, Washington, DC: Federal Highway Administration (GPO). Also ANSI D6.1e-1989.

Martin, D.E. and Kroell, C.K., 1968, Vehicle crush and occupational behavior. *Society of Automotive Engineers Transactions*, vol. 76. Highway Vehicle Safety, SAE, 664–686.

MIL-STD-882D, 2000, *DOD Standard Practice for System Safety*. Washington, DC: Department of Defense (GPO). See www.afmc.wpafb.af. mil/organizations/HQ-AFMC/SE/ssd.htm.

Musculosketetal Disorders and the Workplace: Low Back and Upper Extremities, 2001, National Research Council and Institute of Medicine. Washington, DC: National Academy Press.

Nahum, A.M., Schneider, D.C. and Kroell, C.K., 1975, *Cadaver Skeletal Response to Blunt Thoracic Impact*, paper 751150. Warrendale, PA: Society of Automotive Engineers.

National Research Council, 1983, *Risk Assessment in the Federal Government: Managing the Process*. Washington, DC: GPO.

Nichols, H.L., 1962, *Moving the Earth* (2nd edn) Greenwich, CT: North Castle Books.

Noble, M. and Sanders, A.F., 1980, Searching for traffic signals while engaged in compensatory tracking. *Human Factors*, 22(1), 89–102.

NSPE, 1976, The Engineer will have proper regard for the safety, health, and welfare of the public in the performance of his professional duties. Section 2, *Code of Ethics for Engineers*, NSPE Publication 1102, July 1976, National Society of Professional Engineers, Washington DC. Also contained in *Ethics for Engineers*, NSPE Publication 1105, February 1974.

Ohr, S., 2001, Task simulator equips MSA for real-time response. *Electronic Engineering News*, 23 Apr., 26.

Olson, P.L. and Sivak, M., 1986, Perception–response time to unexpected roadway hazards. *Human Factors*, 28(1), 91–96.

Osborne, D.J., 1986, Vehicle vibration. In Peters, G.A. and Peters, B.J. (eds), 1986 supplement to *Automotive Engineering and Litigation*. New York and London: Garland.

Pachiandi, G., Morgillo, F., Pauzie, A., Deleurence, P. and Giulhon, V., 1994, Impact de l'utilisation du téléphone de voiture sur la sécurité routière (Impact of the use of car phones on road safety). *International Conference on Traffic and Transport Psychology*. *Approche Experimentale sur Simulateur de Conduite* – Rapport LESCO 9417, Oct. 1994. *Approche Experimentale en Conditions Réelles de Circulation* – Rapport LESCO 9510, Oct. 1995. ICTTP 1996 paper TS01.07.

Parkes, A.M., 1991, Drivers' business decision making ability whilst using carphones. In Lovessey, E. (ed.), *Contemporary Ergonomics*, Proceedings of the Ergonomic Society Annual Conference pp. 427–432. London: Taylor and Francis.

Parkes, A.M., 1993, Voice communications in vehicles. In Franzer, S. and Parkes, A. (eds), *Driving Future Vehicles*, pp. 219–228. London: Taylor and Francis.

Patrick, L.M., 1965, Human Tolerance to Impact – Basis for Safety Design, paper 1003B. Warrendale, PA: Society of Automotive Engineers.

Peacock, B. and Karwowski, W. (eds), 1993, *Automotive Ergonomics*. Washington, DC: Taylor & Francis.

Peters, G.A., 1991, Warnings, notices and safety information. In Peters, G.A. and Peters, B.J. (eds), *Sourcebook on Asbestos Diseases*, vol. 5. Charlottesville, VA: Lexis.

Peters, G.A., 1996, Prevention of human failure. In Rossmanith, H.P. (ed.), *Failure and the Law, Proceedings of the Fifth International Conference on Structural Failure, Product Liability, and Technical Insurance*, 10–12 July 1995, Vienna, pp. 5–22. London: Chapman & Hall.

Peters, G.A., 1997, Risk analysis. *Technology, Law and Insurance*, 2, 97–110.

Peters, G.A., 1998a, Airbags as a cause of brain injury. *Technology, Law and Insurance*, 3, 229–236. Also see Peters, G.A., 1999, Airbags as a cause of brain injury. *Hazard Prevention*, Q1, 14–19.

Peters, G.A., 1998b, Product liability and safety, in *The CRC Handbook of Mechanical Engineering*, pp. 20-11–20-15 (F. Kreith ed.). Boca Raton, FL: CRC Press.

Peters, G.A., 1999, Injury reduction in vehicle collisions. *Rx: Law and Medicine*, 1(4), summer, 1, 4–6.

Peters, G.A. and Peters, B.J. (eds), 1984–1993, *Automotive Engineering Litigation*, 6 vols. New York: Wiley.

Peters, G.A. and Peters, B.J., 1999a, *Warnings, instructions, and technical communications*. Tucson, AZ: L&J Publishing Co.

Peters, G.A. and Peters, B.J., 1999b, Occupant injury protection in automobile collisions. *J. R. Soc. Health*, 119(4), 254–260.

Peters, G.A. and Peters, B.J., 2000a, Universal design: A reasonable standard of care. *Journal of System Safety*, Q4, 27–31.

Peters, G.A. and Peters, B.J., 2000b, *Risk Reduction for Future Vehicles*, technical paper 2000-01-3059, in *Intelligent Vehicle Technology*, SP-1558, 21–23 Aug. 2000. Warrendale, PA: Society of Automotive Engineers.

Peters, G.A. and Peters, B.J., 2001, *The Asbestos Legacy*, vol. 23 of the *Sourcebook on Asbestos Diseases*. San Francisco, CA: Matthew Bender Division of Lexis Publishing Co.

Petica, S.,1993, Risks of cellular phone usage in the car and its impact on road safety. *Recherche-Transports-Sécurité*, 37, 45–56.

Petica, S. and Bluet, J.C., 1989, *Rapport sur l'Enquête Internationale Concernant le Radio-téléphone et la Securité Routière*. Rapport INRETS, DOC-126224.

Powell, W.R., Advanti, S.H., Clark, R.N., Ojala, S.J. and Holt, D.J., 1974, *Investigation of Femur Response to Longitudinal Impact*, paper 741190. Warrendale, PA: Society of Automotive Engineers.

Prasad, P. and Mertz, H.J., 1985, *The Position of the United States Delegation to the ISO Working Group 5 on the Use of HIC in the Automotive Environment*, paper 851246. Warrendale, PA: Society of Automotive Engineers.

Raffensberger, C. and Tickner, J. (eds), 1999, *Protecting the Public Health and the Environment: Implementing the Precautionary Principle*. Washington, DC: Island Press.

Redelmeier, D.A and Tibshirani, R.J., 1997, Association between cellular telephone calls and motor vehicle collisions. *New England Journal of Medicine*, 336(2), 453–458.

Red Tractor Book, 37th edn, 1952. Kansas City, MO: Implement and Tractor Publications.

Reimpell, J., Stoll, H. and Betzler, J.W., 2001, *The Automotive Chassis, Engineering Principles*, 2nd edn. Warrendale, PA: Society of Automotive Engineers. Conforms to requirements of SAEJ670 and ISO8855.

Reliability Predictions for Electronic Equipment, 1986, Military Handbook 217. Washington, DC: Government Printing Office.

Risk reduction for future vehicles. SAE 2000-01-3059. In *Intelligent Vehicle Technology*, SP1558. Warrendale, PA: Society of Automotive Engineers.

Road and Paving Materials: Vehicle–Pavement Systems, 2002, vol. 04.03. West Conshohocken, PA: American Society for Testing and Materials.

Robinson, G.H., Erickson, D.J., Thurston, G.L. and Clark, R.L., 1972, Visual search by automobile drivers. *Human Factors*, 14(4), 315–323.

Rosenbluth, W., 2000, *Investigation and Interpretation of Black Box Data in Automobiles: a Guide to the Concepts and Formats of Computer Data in Vehicle Safety and Control Systems*. West Conshohoken, PA: American Society for Testing and Materials and Warrendale, PA: Society of Automotive Engineers.

SAE J885a, *Human Tolerance to Impact Conditions as Related to Motor Vehicle Design*. Warrendale, PA: Society of Automotive Engineers.

SAE Wheel Standards Manual, 2001. Warrendale, PA: Society of Automotive Engineers.

Saran, C., 1988, Agricultural vehicles. In Peters, G.A. and Peters, B.J. (eds), *Automotive Engineering and Litigation*, vol. 2. New York and London: Garland Publishing.

Severy, D.M. 1968, Human simulation for automotive research. In *Highway Vehicle Safety*, pp. 450–481. Warrendale, PA: Society of Automotive Engineers.

Sloan, G.D., 2000, The application of computer simulation and 3-D graphics to forensic human factors. In *Proceedings of the Silicon Valley Ergonomics Conference & Exposition*, San Jose State University, California, pp. 107–118.

Sloan, G.D. and Talbott, J.A., 1996, Forensic application of computer simulation of falls. *Journal of Forensic Sciences*, ASTM, 782–785.

Smith, E.E. and Jonides, J., 1999, Storage and executive processes in the frontal lobes. *Science*, 283, 12 Mar., 1657–1661.

Snyder, R.G., Schneider, L.W., Owings, C.L., Reynolds, H.M. Golomb, D.H. and Schork, M.A., 1997, *Anthropometry of Infants, Children, and Youths to Age 18 for Product Safety Design*, SAE SP-450. Warrendale, PA: Society of Automotive Engineers.

Solomon, S.S., 1999, *Emergency Vehicle Accidents: Prevention and Reconstruction*. Tucson, AZ: Lawyers & Judges Publishing Co.

Spelke, E., Hirst, W. and Neisser, U., 1976, Skills of divided attention. *Cognition*, 4, 215–230.

Stapp, J.P., 1986, Human and chimpanzee tolerance to linear decelerative force. In Sances, A. *et al.* (eds), *Mechanisms of Health and Spine Trauma*. New York: Alloray.

Stein, A.C., Parseghian, Z. and Allen, R.W., 1987, A simulator study of the safety implications of cellular mobile phone use. *31st Annual Proceedings, American Association for Automotive Medicine*, New Orleans.

Stoffer, H., 2000, NHTSA wants to test female crash dummies. *Automotive News*, 31 Jan., p. 16.

Sussman, E.D., Bishop, H., Hadnick, B. and Walter, R., 1985, Driver inattention and highway safety. *Transportation Research Record*, 40–48.

System Safety Handbook, 2nd edn, 2001. Unionville, VA: System Safety Society.

Templer, J., 1992, The Staircase – Studies of Hazards, Falls, and Safer Design. Cambridge, MA: MIT Press.

Triggs, T.J., 1986, Speed estimation. In Peters, G.A. and Peters, B.J. (eds), 1986, Supplement to *Automotive Engineering and Litigation*. New York and London: Garland.

UMTCD, 1988, *Manual on Uniform Traffic Control Devices*, 4D-4. Washington, DC: Federal Highway Administration.

Viano, D.C. and Gadd, C.W., 1975, *Significance of Rate of Onset in Impact Injury Evaluation*, paper 751169. Warrendale, PA: Society of Automotive Engineers.

Violanti, J.M. and Marshall, J.R., 1996, Cellular phones and traffic accidents: an epidemiological approach. *Accident Analysis and Prevention*, 28, 265–270.

Von Alven, W.H., 1964, *Reliability Engineering*. Englewood Cliffs, NJ: Prentice Hall.

Walker, C., 1999, The need for state and worker enforcement of health and safety laws. In Peters, G. and Peters, B. (eds), *Sourcebook on Asbestos Diseases*. Charlottesville, VA: Lexis Division of Reed Elsevier.

Wang, J., Knipling, R.R. and Goodman, M.J., 1996, The role of driver inattention in crashes: new statistics from the 1995 crashworthiness data system (CDS). *40th Annual Proceedings: Association for the Advancement of Automotive Medicine*, pp. 377–392.

Websters New Collegiate Dictionary, 2nd edn, 1956. Springfield, MA: G&C Merriam, p. 789.

Wickens, C.D., Information processing, decision-making, and cognition. In G. Salvendy (ed.), *Handbook of Human Factors*, pp. 72–107. New York: Wiley.

Wierwille, W.W., Casali, J.G. and Repa, B.S., 1983, Driver steering reaction to abrupt-onset crosswinds, as measured in a moving-base driving simulator. *Human Factors*, 25(1), 103–116.

Yoganandan, N., Pintar, F.A., Larson, S.J. and Sances A. (eds), 1998, *Frontiers in Head and Neck Trauma: Clinical and Biomechanical*. Amsterdam: IOS Press.

Zwahlen, H.T., Adams Jr., C.C. and Schwartz, P.J., 1988, Safety aspects of cellular telephones in automobiles, paper 88058, *Proceedings of the ISATA Conference*, Florence, Italy.

Index

T - #0610 - 071024 - C0 - 244/170/11 - PB - 9780367395872 - Gloss Lamination